U.S.-Japan Strategic Alliances in the Semiconductor Industry: Technology Transfer, Competition, and Public Policy

*A15045 261412

Committee on Japan
Office of Japan Affairs

Office of International Affairs
National Research Council

National Academy Press
Washington, D.C. 1992

NOTICE: The project that is the subject of this report was approved by the Governing Board of the National Research Council, whose members are drawn from the councils of the National Academy of Sciences, the National Academy of Engineering, and the Institute of Medicine. The members of the committee responsible for the report were chosen for their special competences and with regard for appropriate balance.

This report has been reviewed by a group other than the authors according to procedures approved by a Report Review Committee consisting of members of the National Academy of Sciences, the National Academy of Engineering, and the Institute of Medicine.

The National Academy of Sciences is a private, nonprofit self-perpetuating society of distinguished scholars engaged in scientific and engineering research, dedicated to the furtherance of science and technology and to their use for the general welfare. Upon the authority of the charter granted to it by the Congress in 1863, the Academy has a mandate that requires it to advise the federal government on scientific and technical matters. Dr. Frank Press is president of the National Academy of Sciences.

The National Academy of Engineering was established in 1964, under the charter of the National Academy of Sciences, as a parallel organization of outstanding engineers. It is autonomous in its administration and in the selection of its members, sharing with the National Academy of Sciences the responsibility for advising the federal government. The National Academy of Engineering also sponsors engineering programs aimed at meeting national needs, encourages education and research, and recognizes the superior achievement of engineers. Dr. Robert M. White is president of the National Academy of Engineering.

The Institute of Medicine was established in 1970 by the National Academy of Sciences to secure the services of eminent members of appropriate professions in the examination of policy matters pertaining to the health of the public. The Institute acts under the responsibility given to the National Academy of Sciences by its congressional charter to be an adviser to the federal government and, upon its own initiative, to identify issues of medical care, research, and education. Dr. Kenneth I. Shine is president of the Institute of Medicine.

The National Research Council was organized by the National Academy of Sciences in 1916 to associate the broad community of science and technology with the Academy's purposes of furthering knowledge and advising the federal government. Functioning in accordance with general policies determined by the Academy, the Council has become the principal operating agency of both the National Academy of Sciences and the National Academy of Engineering in providing services to the government, the public, and the scientific and engineering communities. The Council is administered jointly by both Academies and the Institute of Medicine. Dr. Frank Press and Dr. Robert M. White are chairman and vice-chairman, respectively, of the National Research Council.

This report was prepared with support of a grant from the United States-Japan Foundation. Available from:

Office of Japan Affairs
National Research Council
2101 Constitution Avenue, N.W.
Washington, DC 20418

National Academy Press
2101 Constitution Ave., N.W.
Washington, DC 20418

Library of Congress Catalog Card Number 92-64124
International Standard Book Number 0-309-04779-X
S627

Copyright © 1992 by the National Academy of Sciences

Printed in the United States of America

COMMITTEE ON JAPAN

Erich Bloch, *Chairman*
Council on Competitiveness

C. Fred Bergsten
Institute for International
   Economics

Lewis M. Branscomb
Harvard University

Harold Brown
Center for Strategic and
   and International Studies

Lawrence W. Clarkson
The Boeing Co.

I. M. Destler
University of Maryland

Mildred S. Dresselhaus
Massachusetts Institute of
   Technology

Daniel J. Fink
D. J. Fink Associates, Inc.

Ellen L. Frost
Institute for International
   Economics

Lester C. Krogh
3M Co.

E. Floyd Kvamme
Kleiner Perkins Caufield & Byers

Yoshio Nishi
Hewlett-Packard Co.

Daniel I. Okimoto
Stanford University

John D. Rockefeller IV
United States Senate

Richard J. Samuels
MIT Japan Program

Robert A. Scalapino
University of California, Berkeley

Hubert J. P. Schoemaker
Centocor, Inc.

Ora E. Smith
Illinois Superconductor Corp.

Albert D. Wheelon
Hughes Aircraft Co. (retired)

Ex Officio Members:

Gerald P. Dinneen, Foreign Secretary, National Academy of Engineering

James B. Wyngaarden, Foreign Secretary, National Academy of Sciences
   and Institute of Medicine

SEMICONDUCTOR WORKING GROUP ON PRIVATE SECTOR TECHNOLOGICAL LINKS BETWEEN THE UNITED STATES AND JAPAN

Daniel I. Okimoto (*Co-Chairman*)
Stanford University

Sheridan Tatsuno (*Co-Chairman*)
NeoConcepts

E. Floyd Kvamme
Kleiner Perkins Caufield & Byers

Yoshio Nishi
Hewlett-Packard Co.

Edward J. DeWath
Consultant

OFFICE OF JAPAN AFFAIRS

Since 1985 the National Academy of Sciences and the National Academy of Engineering have engaged in a series of high-level discussions on advanced technology and the international environment with a counterpart group of Japanese scientists, engineers, and industrialists. One outcome of these discussions was a deepened understanding of the importance of promoting a more balanced two-way flow of people and information between the research and development systems in the two countries. Another result was a broader recognition of the need to address the science and technology policy issues increasingly central to a changing U.S.-Japan relationship. In 1987 the National Research Council, the operating arm of both the National Academy of Sciences and the National Academy of Engineering, authorized first-year funding for a new Office of Japan Affairs (OJA). This newest program element of the Office of International Affairs was formally established in the spring of 1988.

The primary objectives of OJA are to provide a resource to the Academy complex and the broader U.S. science and engineering communities for information on Japanese science and technology, to promote better working relationships between the technical communities in the two countries by developing a process of deepened dialogue on issues of mutual concern, and to address policy issues surrounding a changing U.S.-Japan science and technology relationship.

Staff

Martha Caldwell Harris, Director
Thomas Arrison, Research Associate
Maki Fife, Program Assistant

Contents

1. INTRODUCTION 1

2. BACKGROUND: TECHNOLOGY TRANSFER AND
   LATECOMER CATCH-UP 4

3. STRATEGIC ALLIANCES AND TECHNOLOGY
   TRANSFER 9
   Trends in U.S.-Japan Alliances, 11

4. FORCES DRIVING THE FORMATION OF STRATEGIC
   ALLIANCES 15
   Generic Forces, 15
   High-Technology Factors, 17
   U.S.-Japan Alliances, 19
   Business Cycles, Technological Change, and
      Political Factors, 23
   Recessions: Lessons Learned, 25

5. A TYPOLOGY OF ALLIANCES 31
   Direction of Technology Flow: National Interests, 34
   Asymmetrical Pairings: Large and Small Companies, 39
   Symmetrical Pairings: Large Companies, 46
   Why Diversification?, 48

6. ISSUES FOR U.S. POLICY: JAPANESE INVESTMENTS
   AND U.S. COMPETITIVENESS 51

7. ISSUES FOR U.S. POLICY: NATIONAL SECURITY 54

8. ISSUES FOR U.S. POLICY: GLOBAL TECHNOLOGICAL
   STRATIFICATION AND U.S. TECHNOLOGICAL
   CAPABILITIES 59
   Upstream Trends: The Semiconductor Equipment
     Industry, 59
   Downstream Trends: Systems, Components, and
     Proprietary Architectures, 61

9. POSSIBLE SCENARIOS FOR U.S.-JAPAN ALLIANCES
   AND THEIR IMPLICATIONS FOR THE
   UNITED STATES 67
   Scenario 1: Gradual U.S. Recovery, 67
   Scenario 2: Market Share Equilibrium, 68
   Scenario 3: Gradual U.S. Decline, 69
   Scenario 4: Japanese Dominance, 70
   Scenario 5: Pacific Rim Dominance, 71

10. CONCLUSIONS 74
    Examples of U.S.-Japan Alliances: Assessing Costs
      and Benefits, 75
    Semiconductors as a Strategic Industry, 83
    Competitive Advantage: Issues for U.S. Industry, 86
    National Interests: Issues for the U.S. Government, 87

APPENDIXES
A. Case Studies of U.S.-Japan Technology
   Linkages in Semiconductors 91
   Case I: Motorola-Toshiba, 91
   Case II: Sun-Fujitsu, 101
   Case III: Kubota Computer, 109

B. Examples of Japanese Acquisitions and Investments in U.S.
   Semiconductor Companies 113

C. Examples of Japanese Acquisitions and Investments in U.S.
   Semiconductor Equipment and Materials Companies 115

D. Workshop on U.S.-Japan Technology Linkages in Semiconductors:
   Agenda and Participants 117

1

Introduction

Since the birth of the semiconductor industry with the invention of the transistor at Bell Laboratories in 1947, the most significant product and technical advances in microelectronics have been achieved by U.S. companies. Besides the transistor, U.S. companies pioneered the integrated circuit, the dynamic memory, the microprocessor, and other critical products and processes. American companies continue to hold an innovative edge in a number of product areas.

Yet today, the Japanese semiconductor industry is the world market share leader. In 1991, Japanese companies held 46 percent of the $60 billion world market for semiconductors, and U.S. companies held 39 percent. Of the top ten merchant semiconductor companies in the world, six are based in Japan and three in the United States. The current situation contrasts with that of 1970, when Japanese companies held 20 percent of the world market and American companies held 60 percent.

Will the United States continue to lose ground in the semiconductor industry to Japan and other countries? How U.S. semiconductor companies fare carries significance greater than the U.S. industry's $25 billion in sales. Semiconductors are critical components for the nearly $400 billion U.S. electronics industry. Advances in microelectronics will continue to enable systems companies to push the frontiers of information processing and communications, holding out the possibility of new products with widespread impacts—like those of the personal computer during the 1980s. In addition, semiconductors increasingly allow companies to build more sophisticated functions into nonelectronic products such as automobiles. The role that microelectronic devices play in high-technology weapons systems—already significant—will likely increase as well.

These issues form the context for this assessment of U.S.-Japan strategic alliances in the semiconductor industry, which focuses on alliances that transfer or develop technology. The purpose of the study is to examine the scope and nature of alliances, to identify the forces behind them, and to consider the impacts on the participating organizations and the United States as a country.

Strategic alliances have long played a limited role in the semiconductor industry. By the time the Japanese industry had established its leading position in the mid-1980s, however, it was clear that the number of U.S.-Japan alliances was increasing sharply from the low level of activity prior to 1980. This general trend has continued to the present. Further, U.S.-Japan strategic alliances appear to have become deeper and more significant in terms of their impact on companies and on the semiconductor industry's competitive landscape.

To summarize some of the major themes of this report, a number of forces have contributed to the expansion of U.S.-Japan alliances. From the U.S. perspective, small start-up firms increasingly turn to large Japanese companies to supplement or replace traditional sources of financing for growth, such as venture capital. For small U.S. firms and large American companies, linking with Japanese partners can provide access to advanced manufacturing capability and to the rapidly growing Japanese market—now the largest in the world. For the large, integrated "silicon majors" that dominate Japanese industry, linkages with small U.S. firms provide access to complementary technical capabilities that can be leveraged to gain a stronger position in new, design-intensive semiconductor markets as well as downstream systems. In the mid- and late 1980s the majors were joined in forming U.S. alliances by "lateral entrants"—large Japanese steel and equipment companies that used linkages with U.S. companies to acquire the critical mass of technology necessary to diversify into semiconductors and other information industry markets.

Extensive U.S.-Japan alliance activity in an environment of fierce bilateral competition is here to stay for the foreseeable future. When structured properly, alliances can bring substantial short-term benefits to both sides. Because of a more favorable environment for intellectual property protection, some small U.S. companies have been able to structure better alliances than they could in years past, and U.S. access to the Japanese market has been facilitated by alliances.

Yet the spread of alliances raises concerns as well. This report documents that the prevailing flow of semiconductor technology through alliances is from the United States to Japan. Growing U.S.-Japan technical interdependence in semiconductors may reinforce structural weaknesses—particularly on the U.S. side—that lead to an imbalance in long-term benefits. A largely one-sided outflow of technology from the United States to Japan, if continued over the 1990s, could have the cumulative effect of eroding the foundations of America's capacity to innovate in this industry. This erosion would have serious consequences for U.S. computer and telecommunications companies that use semiconductors, for the overall U.S. economy as we move into the information age, and for national security.

For strategic alliances to bring balanced, long-term benefits to participants and to the United States and Japan as countries, it will be necessary to redress these structural weaknesses, which include manufacturing and process technology in the United States, and generic research and new product design in Japan. For this to happen, significant changes will be required on both sides. There are increasing opportunities for U.S. companies that have built the necessary capabilities to access Japanese technology, as well as some encouraging signs that actions on the part of both industry and government are strengthening the manufacturing infrastructure for U.S. industry. Despite these encouraging signs, however, the larger competitive calculus of an expanding Japanese global market share persists.

New trends in the computer industry and other downstream sectors that buy semiconductors—such as global consortia built around microprocessor standards—will heighten competition in the years to come. The ascendance of companies in South Korea and elsewhere in the most capital-intensive segments of the semiconductor industry will create new challenges for U.S. and Japanese companies.

The critical question for individual U.S. companies and industry leaders is how to build and implement strategies for maximizing the benefits of alliances with Japan so that the United States remains a front-line player in all aspects of the semiconductor industry, from basic research and design to manufacturing and marketing. The main issue for the U.S. government is to adopt policies favorable to U.S. industry strategy building and long-term competitiveness.

This report was prepared by a working group of experts as part of a project initiated by the National Research Council's Committee on Japan to examine technology linkages between Japan and the United States. Co-chaired by Daniel Okimoto of Stanford and Sheridan Tatsuno of NeoConcepts, the working group was formed in the fall of 1990 and met a number of times in 1991 to deliberate and confer on analysis and data collection. A workshop on U.S.-Japan Technology Linkages in Semiconductors was convened in September 1991 to gain additional insights from other experts in the United States and Japan. The staff of the National Research Council's Office of Japan Affairs, which also serves as the staff for the Committee on Japan, assisted the working group in data collection, and in analysis and compilation of results.

2

Background: Technology Transfer and Latecomer Catch-up

Japan has reaped the incalculable benefits of a steady stream of scientific and technological know-how from the United States and Europe.[1] From 1951 to 1984—a time frame encompassing Japan's postwar economic reconstruction and consolidation during the 1950s, rapid growth during the 1960s, adjustment to oil shocks of the 1970s, and breakthrough to prominence in the semiconductor and other high-technology industries during the early 1980s—Japanese corporations concluded more than 40,000 contracts with foreign firms, providing for the transfer of technologies deemed critical for commercial competitiveness in domestic and world markets.[2] This windfall of foreign technology included such seminal patents as Du Pont's for nylon (used in synthetic textiles), RCA's for basic color television technology (for television and consumer electronics), and transistors from Bell Laboratories (for integrated circuits and the information industries).

Having access to seminal technologies from abroad—adapted and upgraded by Japanese importers—paved the way for Japan to emerge as a world-class manufacturing power. Indeed, most analyses of Japan's fast-paced growth into economic superpower status give heavy weight to the

---

[1] Terutomo Ozawa, *Japan's Technological Challenge to the West, 1950-1974: Motivation and Accomplishment* (Cambridge: MIT Press, 1974).

[2] James C. Abegglen and George Stalk, Jr., *Kaisha: The Japanese Corporation* (New York: Basic Books, 1985), pp. 126-127.

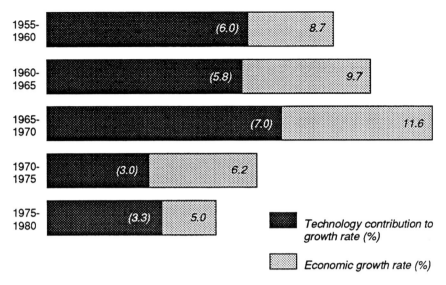

FIGURE 1 The Japanese challenge in high technology. SOURCE: Dr. Hisao Kanamori, from Daniel I. Okimoto "The Japanese Challenge in High Technology," Ralph Landau and Nathan Rosenberg, eds., *The Positive Sum Strategy: Harnessing Technology for Economic Growth* (Washington, D.C.: National Academy Press, 1986).

contributions made by technological improvements.[3] According to one prominent Japanese economist, technology (much of it imported) was responsible for more than half of Japan's economic growth between 1955 and 1980 (see Figure 1). Other nations and industries—such as the U.S. shipbuilding industry in the eighteenth century—have implemented "catch-up" strategies in which the transfer of technology from abroad was a major element. What is significant about Japan's experience is the systematic, organized way in which technology has been imported.

In return for billions of revenue dollars generated by the transfer and adaptation of foreign technology, Japanese companies paid a relatively modest cumulative sum of only $17 billion. Amortized over 33 years, Japanese industry paid, on average, only about $500 million per year, a fraction of what it undoubtedly would have cost to develop the technology at home, provided Japanese companies could have achieved the breakthroughs. For

---

[3]Edward F. Denison and William K. Chung, "Economic Growth and Its Source" in Hugh Patrick and Henry Rosovsky, eds., *Asia's New Giant: How the Japanese Economy Works* (Washington, D.C.: The Brookings Institution, 1976), especially pp. 125-130.

individual firms struggling to be competitive, the royalty payments proved burdensome, taking a significant bite out of revenues and limiting the earnings that could be ploughed back into research and development. However, royalty payments represent only a portion of the up-front costs and risks of R&D incurred by foreign patent holders—to say nothing of the uncertainties, false starts, and time required for the processes of invention. For Japanese firms, the benefits of having access to foreign technology far outweighed the marginal costs.

It would be hard to exaggerate the advantages of being in a position to buy foreign technologies "off the shelf." With modifications, leading-edge technologies could be put to immediate use in manufacturing. For Japanese companies, the immense benefits included crucial time saved, large uncertainties eliminated, promising R&D pathways clarified, rapid movement down technological and commercial learning curves, resources freed to focus on incremental adaptations, and new commercial opportunities opened up. Without the infusion of key foreign technology, Japanese industry probably would have advanced less rapidly and not as synergistically across so many fronts.[4]

At the same time, it must be noted that the most significant transfers of technology occurred many years ago. Experts may disagree about the point at which U.S. companies had transferred the critical mass of technology to companies in Japan, but the time is probably 10 to 20 years ago. Japanese companies have invested considerable resources in adapting, commercializing, and further developing these technologies. They are now global technological leaders in many segments of this industry, and the high rate of current R&D investments by these companies indicates their long-term viability as formidable competitors.

From the standpoint of U.S. companies that sold advanced technology, the consequences were not as positive. Individual U.S. firms may have earned a steady stream of patent revenues, bolstering quarterly dividends and yearly profits, but most U.S. companies failed at the time to appreciate the impetus for rapid catch-up that the transfer of technology gave to competitor firms in Japan. For the U.S. economy as a whole, the long-term effects of technology bartering were debilitating to the extent that it contributed to an erosion of America's industrial preeminence. James Abegglen and George Stalk go so far as to call the one-way outflow "disastrous."[5]

---

[4]Christopher Freeman stresses the multiplier-effect benefits for Japan of imported technology, including especially "reverse engineering." See Christopher Freeman, *Technology Policy and Economic Performance: Lessons from Japan* (London: Pinter Publishers, 1987), pp. 39-49.

[5]Abegglen and Stalk, op. cit., p. 128.

Reflecting on the circumstances at the time, however, we can understand that U.S. companies had reasons for selling their hard-earned technology. Many were preoccupied with the huge and expanding U.S. domestic market, one so much bigger and more enticing than any other national market. Because success in the domestic market would bring in profits that would dwarf anything earned from foreign markets, why bother incurring the costs, risks, and uncertainties of trying to break into what was, in the 1960s, a small and distant Japanese market? Why not sell patents to bolster profits? Patent sales required little or no up-front investment of time or money, and few people expected Japanese companies to transform themselves soon into world-class competitors. The idea of second sourcing in Japan also had great appeal.

What about far-sighted U.S. companies, such as Texas Instruments, Motorola, Intel, and National Semiconductor, which understood the long-term importance of breaking into the Japanese market? These companies ran into the roadblock of formal and informal barriers as they tried to enter the Japanese market, which led them to abandon or delay early plans to establish a presence in Japan. Under such circumstances, they came to the conclusion that earning royalties from the sale of their technology patents was better than having nothing to show for their efforts to penetrate the Japanese market. National Semiconductor, an unusual case, was able to set up operations in Japan only by purchasing a plant in Okinawa shortly before reversion of the islands to Japan.

American companies have learned some hard lessons from past experience. Today, the value of state-of-the-art technology is recognized more clearly than it was two decades ago. American executives realize that the possession of key technologies can be converted into major gains in the commercial marketplace. Accordingly, many U.S. companies in the semiconductor industry have tightened up their licensing practices, often choosing either not to license at all or to use their patents to obtain know-how of comparable value in return.

Even if companies take a more cautious approach to licensing their technology, there are many other avenues through which know-how can travel. For basic research, the channels of transmission include published articles and books, conference papers and public discussions, graduate training, contract research, consulting services, and corporate participation in university-based laboratory research. For product and process technology, the channels include reverse engineering of already manufactured products, industrial espionage, consulting services, hiring away of key researchers from competitor companies, and various forms of strategic alliances. For companies in industries such as semiconductor manufacturing equipment, the sale of a product inherently involves technology transfer to the customer.

Closing off the flow of knowledge in basic research is not only impossible but also undesirable. Carefully structuring the transfer of product and process technology, on the other hand, is not only desirable but also more feasible. No matter how much the faucets are tightened, technology will continue to flow out. Indeed, the semiconductor industry is today inherently global—no company (U.S. or foreign based) is completely self-sufficient in terms of technology.

Japanese companies are still importing technologies developed in the United States, such as reduced instruction set computing (RISC), parallelism, and video compression. At the same time, there is now evidence of export of technologies developed in Japan. Intel has a strong position in flash memories and will shore up its market position through an alliance with Sharp. Although Toshiba published early work on flash memories, it was Intel that developed the first application of the technology that was widely successful in the market. Semiconductor companies cannot survive without global technology linkages. The question today, then, is not how these linkages can be reduced or avoided, but how to make them work best for participating U.S. companies and for the United States as a country.[6]

America's early postwar experience with a continued outflow of technology has led to an outpouring of concerns about its corrosive long-term effects on the competitiveness of U.S. companies in the high-technology sectors and on the U.S. economy as a whole.[7] Measures to slow or control the outflow of technology have been prescribed, including stricter monitoring and restriction of foreign direct investments and the imposition of tighter reins on certain types of strategic alliances. Arguments in favor of tighter controls clash with the deeply entrenched ideology of free market economics and the presumed advantages of security alliance structures (especially the U.S.-Japan alliance). Although everyone wants to protect U.S. national interests, analysts disagree on how this might best be accomplished. This report seeks to highlight some of the problems and some of the opportunities associated with U.S.-Japan alliances for practitioners and policymakers who must consider how to make these rapidly expanding relationships demonstrate concrete benefits for the United States.

---

[6]For purposes of this report, a "U.S. company" is one in which more than half of the equity is held by U.S. citizens. This approach is convenient but not totally satisfactory because some foreign-owned companies may contribute more to the U.S. industrial and technology base than U.S.-owned companies. See Robert Reich, "Who is Us?" *Harvard Business Review*, January-February 1990, pp. 53-64.

[7]See, for example, Linda Spencer, *Foreign Investment in the United States: Unencumbered Access* (Washington, D.C.: Economic Strategy Institute, May 1991).

3

Strategic Alliances and Technology Transfer

For semiconductor companies shopping for external technology, the mechanisms of acquisition range from OEM (original equipment manufacturing) licensing, at the simple end of the spectrum, to joint ventures and the outright purchase of innovative companies, at the more complex and expensive end. Virtually every linkage imaginable along this spectrum falls into the catchall category of "strategic alliances." Indeed, the category is so encompassing that it runs the risk of failing to reveal meaningful distinctions.

For purposes of this report, all alliance linkages that do not serve as a direct conduit for knowledge transmission can be omitted. Specifically, this means the exclusion of strategic alliances dealing exclusively with pure marketing, short-term, spot-market buyer-seller transactions, and arms distance equity investments (made solely to extract financial profits, with no attempt to monitor or obtain technology). Depending on the nature of the exchanges, some instances of these transactions may lead to a transfer of technology. For example, marketing and sales agreements that involve after-service can diffuse technology as an unintended by-product. Nevertheless, certain types of strategic alliances—particularly the simple, one-shot transaction variety—are omitted from consideration here.

A wide range of linkages that involve technology transfer still remains. Dataquest has broken American-Japanese semiconductor alliances down into 12 types. These alliance types can be modified, expanded, and grouped under four different headings, each dealing with different phases of business activity: (1) research and development, (2) manufacturing, (3) marketing and services, and (4) general purpose (see Table 1).

Categorizing alliance types can help us analyze their structure and effects. However, areas of ambiguity remain that defy clarification and precise categorization. In the case of product- or technology-oriented equity

TABLE 1 Typology of Alliances

R&D ALLIANCES
1. *Licensing agreement*: legal permission to utilize patents or proprietary technology for an up-front fee and/or royalties.
2. *Cross-licensing agreement*: two or more companies give legal permission to use each other's patents or proprietary technology.
3. *Technology exchange*: a swap of proprietary technologies, which may or may not involve a transfer of money.
4. *Visitation and research participation*: the dispatch of researchers to visit, observe, and participate in the R&D activities of partner firms.
5. *Personnel exchange*: an ongoing and reciprocal program in which researchers from one company spend time working at the partner company.
6. *Joint development*: two or more companies joining forces to develop new products or technology.
7. *Technology acquisition investments*: foreign investments in companies aimed at gaining access to technology, especially in small, start-up or innovative, medium-sized firms.

MANUFACTURING ALLIANCES
8. *Original equipment manufacturing (OEM)*: manufacturing a product for another company, which sticks its label on it and handles all aspects of business activities, including marketing and servicing, as if it had manufactured the product itself.
9. *Second sourcing*: an arrangement whereby a company is given permission to manufacture a product designed and developed by another company as a second source of supply for customers, using the same specifications.
10. *Fabrication agreement*: use of another company's fabrication facilities to manufacture a product (because the partner lacks its own manufacturing facilities or wishes to subcontract out the task of fabrication).
11. *Assembly and testing agreement*: components and parts manufactured elsewhere are sent to another company where they are assembled and tested.

MARKETING AND SERVICE ALLIANCES
12. *Procurement agreement*: a commitment to purchase certain quantities of specific goods or services over a specified period of time.
13. *Sales agency agreement*: exclusive or nonexclusive rights to sell the partner's original products, or products to which value is added, in specified markets.
14. *Servicing contracts*: the provision of follow-up service in foreign markets (often tied to marketing arrangements).

GENERAL-PURPOSE TIE-UPS
15. *Standards coordination*: an agreement on common or compatible technical standards linking devices and systems and users of different machines.
16. *Joint venture*: two or more firms jointly form a company to develop, manufacture, or market new products.

SOURCE: For Table 1 and unless otherwise indicated, prepared by NRC working group.

investments, for example, boundary questions arise with respect to the level of capital invested. If the percentage of investment is significant (40 or 50 percent), some might call it a strategic alliance. Conversely, if the percentage falls below a certain threshold (25 percent for example), some might conclude (incorrectly) that the level of investment is too trivial to qualify for inclusion as a strategic alliance. Because the main focus of this report is on the transfer of technology, the establishment of an arbitrary capital investment cut-off point—whether 25 percent or 33 percent—is irrelevant and potentially misleading. Key technologies can be obtained with only 5 percent equity investment; conversely, technology might not be transferred even when there is majority ownership. Arbitrary thresholds cannot be established as meaningful guidelines.

This analysis of alliance types focuses on those strategic alliances that serve as conduits for the transmission of technology in one or more directions, either on a one-time basis or as a mechanism for repeated and regular transmission. It would, of course, be useful if the various types of alliances could be correlated with the commercial importance of the technology transferred but unfortunately no discernible pattern of correlation emerges from the data. Thus, we must be content with an operational definition of strategic alliances based on the construction of a typology that highlights the *nature* (iterative or non-iterative) and *direction* (one- or two-way flow) of technology transfer.

TRENDS IN U.S.-JAPAN ALLIANCES

Over the postwar period the scope and nature of American-Japanese alliances have changed, reflecting changes in technology, commercial competition, and the positions of the U.S. and Japanese semiconductor industries. In the earliest phase, covering roughly three decades from 1950 to 1980, very few strategic alliances were concluded. The few that were consummated took the form of simple licensing agreements, involving the sale of basic U.S. patents to latecomer companies in Japan. Until the 1970s, most American-Japanese alliances fell into the category of licensing agreements aimed at transferring basic R&D know-how to Japanese semiconductor producers.

Among the notable licensing agreements were RCA-Hitachi (1961), TRW-Mitsubishi (1962), Honeywell and Fairchild with NEC (1962), Sperry Rand-Oki (1963), and General Electric-Toshiba (1964). Such alliances occurred at a snail's pace of roughly one per year. There were almost no alliances in the areas of manufacturing, marketing and servicing, or general-purpose cooperation, three of the four alliance categories mentioned earlier.

Those U.S. companies wishing to pry open the Japanese market ran squarely into the dead end of formal barriers to trade and foreign direct investment. Unable to export or set up shop, several U.S. firms decided, as a fallback position, to license their technology so as to gain at least some revenue stream. Others, such as Texas Instruments, which were determined to establish a foothold in Japan, had to agree, as the price of admission, not only to license their basic technology but also to enter a joint venture with a domestic producer. Texas Instruments' joint venture with Sony Corporation was dissolved as soon as the American partner was allowed by the Ministry of International Trade and Industry (MITI) to establish a wholly owned Japanese subsidiary.[8] Hence, the alliance pattern for the first 30 years consisted of a few scattered licensing agreements, which grew in numbers during the 1970s; several joint ventures (including the Texas Instruments-Sony venture encouraged by MITI and the unsuccessful Motorola-Alps venture described in the Motorola-Toshiba case study in Appendix A); and very little else.

One explanation for the paucity of alliances relates to the legal limits placed on foreign direct investments in Japan, the strict regulatory controls imposed over foreign exchange, and the requirement that Japanese corporations secure formal MITI approval for licensing and alliance formation.[9] Most of these legal and regulatory constraints were not fully removed until around 1980. The passage of the Foreign Trade and Exchange Act in 1979 cleared the way for Japanese companies to enter freely into joint ventures (with only the obligation of prior notification), to move yen and dollars freely in and out of Japan, and to buy and barter technology without government approval.

This sweeping away of administrative obstacles also happened to occur at a time of very rapid development in the history of Japan's semiconductor industry. The co-occurrence of the two, along with other underlying forces at work, created the circumstances for an explosion of alliances from the mid-1980s onward. During the early 1980s, the frequency of alliance formation increased markedly, especially in the area of U.S. licensing of memory and microprocessor technology. In the mid-1980s, a number of agreements were signed in the area of semiconductor equipment, and by the late 1980s, the proliferation of U.S.-Japan alliances had reached a peak. Just as significantly, the nature of alliances underwent a change from simple licensing to more complicated and multifaceted arrangements (see Figure 2).

Aggregate data available from Dataquest and the American Electronics Association Japan Office confirm the boom in the number, type, and scope of strategic alliances involving U.S. and Japanese semiconductor companies

---

[8] For an account of the Texas Instruments-Sony relationship, see Akio Morita, *Made in Japan* (New York: Weatherhill, 1987), pp. 187-188.

[9] Daniel I. Okimoto, "Outsider Trading: Coping with Japanese Industrial Organization," in Kenneth Pyle, ed., *The Trade Crisis: How Will Japan Respond?* (Seattle: Society for Japanese Studies, 1987), pp. 85-116.

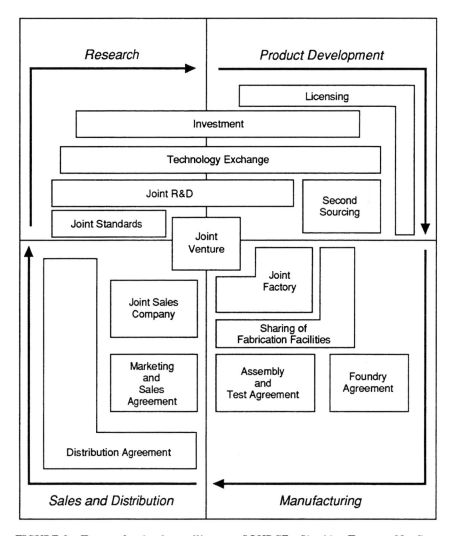

FIGURE 2 Types of technology alliances. SOURCE: Sheridan Tatsuno, NeoConcepts, 1990.

(see Figure 3). Bear in mind that the aggregate data are incomplete because they cover only the strategic alliances that companies have made public. The number of publicly announced alliances may be only the tip of the iceberg, because many alliances are not publicly announced and do not enter the realm of public knowledge. For a variety of reasons, such as fears of sending the wrong signals to investors or of setting off strong competitive counterstrategies on the part of rival firms, American and Japanese companies often choose to keep their alliances quiet.

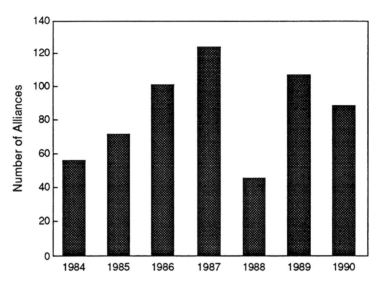

FIGURE 3 American-Japanese semiconductor alliances: 1984-1990. SOURCE: Bruce Kogut and Dong-Jae Kim, "Strategic Alliances of Semiconductor Firms," unpublished report to Dataquest, January 1991.

Therefore, it is almost impossible to estimate the total number of U.S.-Japan alliances with precision. Interviews with U.S. executives suggest that the available figures greatly understate the actual number; indeed, public data may represent less than half of the actual number of American-Japanese alliances. Although data are incomplete, it is clear that U.S.-Japan alliance activity in semiconductors has generally followed an upward trend over the past decade. That trend is not likely to fall off sharply.

One obvious factor explains the proliferation of alliances from the mid- to late 1980s: the growing and immense size of the Japanese semiconductor market, which had become the largest in the world by 1988. Seeing this trend clearly, U.S. semiconductor firms realized their need to find a way to ride the crest of Japanese growth. Strategic alliances with Japanese partners offered a vehicle for gaining a foothold in Japan's growing but, for many, still difficult to penetrate market. Conversely, Japan's rapidly growing semiconductor industry absorbed large amounts of foreign technology through alliances that might not have been developed by U.S. partners relying on their own resources. Joint ventures, marketing, sales, distribution, servicing, and standards coordination—relatively recent alliance types—can be traced to the dramatic growth of the Japanese market and the commercial imperative for U.S. companies to get involved.

4

Forces Driving the Formation of Strategic Alliances

GENERIC FORCES

Why has the number of strategic alliances risen so dramatically over the past five or six years? To understand the reasons requires that the semiconductor experience be placed in a global context. From a global perspective, the semiconductor industry reflects an emerging pattern of closer international interdependence. The numbers of U.S.-Japan linkages have multiplied across nearly all trade-related sectors, including old-line manufacturing (e.g., automobiles), the services (e.g., banking and securities), and high technology (e.g., biotechnology and aircraft). As discussed in more detail below, an alliance boom in semiconductors occurred in the mid-1980s and again in the late 1980s. Within the past year or so, however, it appears that the number of newly formed U.S.-Japan alliances may be declining somewhat while U.S.-U.S. alliances are expanding. Data on venture capital investments from Japan provide some evidence but do not give a full picture of trends in other areas such as acquisitions.[10] Still, today's level of activity remains far above the early 1980s.

---

[10]Venture Economics reports that in 1990 the Japanese invested $23 million in the U.S. semiconductor industry, while in 1991 the level of investment dropped to $11.25 million (communication with Venture Economics, March 1992). It is also important to note that a downturn in U.S.-Japan alliances is occurring in the context of the growth of global consortia and standards-based groupings, as discussed in more detail later.

The increase in semiconductor alliances since 1985 requires explanation, because the semiconductor industry saw perhaps the highest level of activity across all industrial sectors (with the possible exception of biotechnology). What generic forces are at work propelling companies to forge alliances? What industry-specific factors have led to the proliferation of U.S.-Japan linkages in the semiconductor industry?

Perhaps the most obvious generic force is a trend toward the "globalization of markets."[11] Companies cannot afford to confine themselves to domestic markets, no matter how large they may be. Instead, they must compete in all major markets around the world or risk falling out of long-run contention. For most products, mass volume sales, involving global markets, constitute the *sine qua non* of low-cost production. Those companies failing to compete worldwide will lose the advantages of rapid movement down steep learning curves.

If the world market cannot be partitioned into national units and if semiconductor producers aspire to survive in the crucible of world competition, it is essential that they find ways of getting close to foreign customers. Because the up-front costs and risks of breaking into foreign markets can be prohibitively high, there are strong incentives for companies to find foreign partners to distribute and sell their products through marketing agreements, one form of strategic alliance.

To compete effectively in foreign markets over the long haul, however, companies must establish their own physical presence abroad so that they can understand and respond to local needs. A physical presence makes it possible for companies to do what is necessary to succeed: communicate continually with local customers; learn how to operate in foreign environments; cultivate long-term relationships; develop, adapt, or custom-make products for foreign end users; manufacture locally; and offer timely delivery and reliable after-service. Doing this from a distance is impossible.

Very large corporations can project a physical presence in foreign markets by going multinational,[12] this is what IBM, AT&T, NEC, Toshiba, Hitachi, Texas Instruments, Motorola, and others have done. They have built wholly owned subsidiaries that market, design, manufacture, and provide service for customers in their own local markets. However, because multinationalization of this type is beyond the reach of smaller companies, their only alternative is to develop a series of strategic alliances—such as marketing agreements and manufacturing licenses—that serve the same functional purpose. Such ambitious and fast-developing companies as Sun

---

[11] Kenichi Ohmae, *The Borderless World: Power and Strategy in the Interlinked Economy* (New York: Harper Business, 1990).

[12] Christopher A. Bartlett and Sumantra Ghoshal, *Managing Across Borders: The Transnational Solution* (Boston: Harvard Business School Press, 1989).

Microsystems and MIPS have concluded a series of strategic alliances that have positioned them to carve out significant shares of Japanese and European markets. Even for very large firms the costs and risks of having to compete around the world have become so high that they actively seek alliances not only for themselves but also for their multinational enterprises. IBM Japan, for example, has put in place an extended network of alliances with nearly a thousand local Japanese companies, which has helped it achieve a strong "insider" position in the Japanese market.[13]

HIGH-TECHNOLOGY FACTORS

The heavy concentration of U.S.-Japan alliances in the high-technology sectors, especially computers, biotechnology, and semiconductors, can be linked to the special characteristics of technology-intensive industries. The rapid development of technology, wide scope for ongoing innovation, and continual coming on stream of new generations of products lead to short product life cycles and very high risks, which give rise, in turn, to cost- and risk-reducing alliances. The drive for innovation means that commercial competition is bound to be fierce and that the incentives to enter into strategic alliances are apt to be powerful. In semiconductors as in computers and biotechnology, no single company can dominate in all product markets; few, indeed, can even realistically hope to be active across a spectrum of markets.

Given the need for continuous innovation, most high-technology companies build strategic alliances in order to: (1) compensate for in-house weaknesses or technological gaps; (2) fill out product lines and portfolios; (3) position the company to enter lucrative new markets; (4) better serve an established or targeted customer base; and (5) reduce the costs, risks, and time required to develop new products and process technologies. To achieve these goals, U.S. and Japanese companies are willing to swap technology (e.g., cross-licensing), second source, undertake joint development projects, and organize joint ventures. In more traditional fields, such as textiles and steel, where the technology is more mature and the time required for one generation of new products to supplant another is far longer, the incentives for companies to enter into strategic alliances are much less powerful (unless such alliances permit them to branch out in new directions).

Semiconductor and computer companies find themselves trapped in the squeeze between the spiraling costs and risks of R&D and continuous capital investments, on the one hand, and the collapsing time intervals during which profits on current-generation products can be made, on the other.

---

[13]Ohmae, op. cit., p. 131.

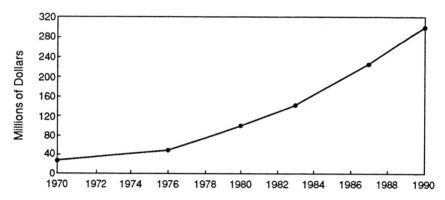

FIGURE 4 Building plus equipment costs for a high-volume fabrication line. SOURCE: Dataquest, April 1991.

The fixed costs of doing research and of building new plant facilities for the 16-megabit DRAM, for example, have soared to almost prohibitive heights—even for giants in the electronics industry (see Figure 4). Such costs would be tolerable if the time intervals within which companies could earn satisfactory profits were long enough; however, the contraction in product life cycles means that the window of opportunity for rent retrieval is exceedingly short—usually only the first year or so of a new product's introduction. Although there were widespread reports in early 1992 of plans by Japanese semiconductor companies to cut capital spending and perhaps R&D spending, the top five Japanese merchant semiconductor companies have consistently outspent the top five U.S. merchants in R&D in recent years.[14]

For dynamic random-access memory (DRAM) and other commodity very large-scale integration (VLSI) chips, perhaps the key competitive requirement is achieving economies of scale. Semiconductor companies must be able to cross a high production threshold to justify the costs of R&D and new plant investments. If companies want to stay competitive in the VLSI business, they must be able to pay for continuous increases in R&D and manufacturing investments at a rate of around 20 to 30 percent per year. For many companies, especially merchant houses, the escalating financial burdens are too heavy to bear. Accordingly, the incentives to enter into manufacturing alliances with Japanese firms are strong. The only other options for VLSI companies are unattractive ones: raising large sums for investment financing or ultimately having to withdraw from DRAM and commodity chip production.

---

[14]National Advisory Committee on Semiconductors, *Attaining Preeminence in Semiconductors*, February, 1992, p. 23. This chart is based on Dataquest figures.

The continuous waves of innovation—the constant coming on stream of new products and of diverse, new families of products—mean that the financial, technological, and manpower resources of even the largest companies are bound to be stretched to their limits. Even highly diversified companies, possessing deep financial pockets, cannot escape the discomfort of being stretched in too many directions. Hence, the imperatives of commercial competition have forced semiconductor producers to be receptive to the idea of linking up with foreign partners, because alliances offer companies the opportunity to pool technological and manpower resources. To cite a recent example: the world's largest computer manufacturer, IBM, has decided to enter into a joint venture with Siemens to manufacture 16-megabit DRAMs in France. IBM has been actively forming alliances with small companies in Japan and Europe, as well as the United States, investing more than $1 billion in 200 companies over the past 10 years.[15]

So compelling have corporate incentives become that crisscrossing linkages of strategic alliances now tie together virtually all major (and most minor) semiconductor companies in ever denser and more complicated networks. One would be hard-pressed to find a single significant semiconductor company anywhere in the world today that has managed to remain isolated from the powerful pull of alliance linkage. Compare the current situation with the 1960s and 1970s, when strategic alliances were the exception rather than the rule, and the dimensions of the recent alliance explosion can be appreciated fully.

U.S.-JAPAN ALLIANCES

If generic and semiconductor-specific forces explain the ebb and flow of alliance formation, what explains the marriage of U.S. and Japanese firms? Why have companies from the two sides of the Pacific joined together in so many alliances—conspicuously more than American-European or Japanese-European cases? According to one study, European investments in U.S. high technology are rather limited (about 16 percent for all industries and about 5 percent of the foreign investments in semiconductors)[16] (see Figure 5). According to the same study, Japanese investments in semiconductors make up about 90 percent of all foreign investments in

---

[15]"Learning from Japan," *Business Week*, January 27, 1992, p. 55.
[16]Linda Spencer, *Foreign Investment in the United States: Unencumbered Access* (Washington, D.C.: Economic Strategy Institute, 1991). The ESI data base includes a variety of print and electronic sources updated periodically. Other sources of data on foreign investments in U.S. high-technology industries include the American Electronics Association Tokyo Office, Ullmer Brothers, Venture Economics, and Dataquest. Dataquest compiles information on many aspects of the electronics industry, including investments, licensing, and other tie-ups that do not involve equity purchases.

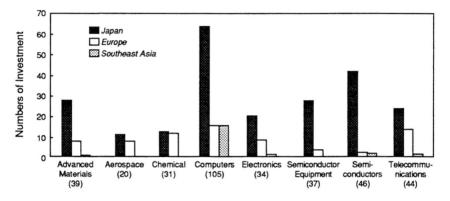

FIGURE 5 Foreign investments in U.S. high technology. NOTE: Numbers in parentheses refer to industry totals. SOURCE: Adapted from Linda Spencer, *Foreign Investments in the United States: Unencumbered Access* (Washington, D.C.: Economic Strategy Institute) May 1991, p. 10.

this industry and about 60 percent of all foreign investments in U.S. high-technology industries. The most obvious explanation for the preponderance of U.S.-Japan alliances in semiconductors is that the U.S. and Japanese semiconductor industries are the biggest and best developed in the world. In 1991, Japan accounted for 38 percent of the world market, and North America 29 percent (see Figure 6). Japanese companies control 46 percent and North American companies 39 percent of total semiconductor sales (see Figure 7). The extraordinary level of economic integration and high degree of trade interdependence are background factors leading to alliances between U.S. and Japanese companies—not only in semiconductors but also across a broad spectrum of industrial sectors.

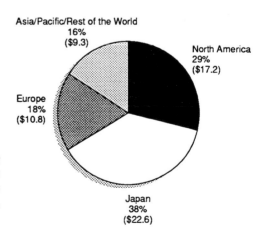

FIGURE 6 World semiconductor market in billions of dollars. NOTE: Total greater than 100% due to rounding. SOURCE: Dataquest, June 1992.

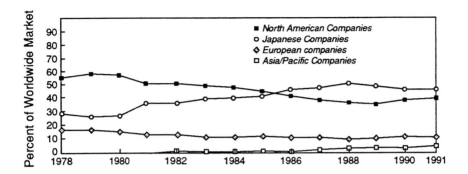

FIGURE 7 Regional shares of the worldwide semiconductor market. SOURCE: Dataquest, June, 1992.

Differences in national systems of industrial organization (levels of vertical integration and diversification within firms, corporate finance, and capital markets) have brought opposites together[17] (see Table 2). Small, venture start-up firms in the United States, in need of funds, manufacturing foundries, and marketing outlets, look to large, deep-pocketed, vertically integrated, and diversified Japanese corporations to meet these needs, since large diversified U.S. corporations have shown far less interest until quite recently in forming alliances with small U.S. venture start-up companies. Many of the large Japanese giants, in turn, look to small U.S. start-ups to provide new product designs to fill niche markets or to compensate for certain deficiencies in their own innovative capabilities. Since the largest returns on alliance investments—such as the development of new technologies or company positioning for long-term diversification into whole new fields—are often reaped only after a long period of gestation, it is not surprising that Japanese companies (with their long time horizons) are among the most active in alliance partnerships.[18] In the best of circumstances, opposite firms attract and combine in ways that overcome the respective limitations of different industrial systems.

Although there are signs of change, it is true that during the past decade, large U.S. corporations, faced with the same opportunities as Japanese

---

[17] Daniel I. Okimoto, Takuo Sugano, and Franklin Weinstein, eds., *Competitive Edge: The Semiconductor Industry in the U.S. and Japan* (Stanford, Calif.: Stanford University Press, 1984).

[18] Many large U.S. corporations diversify into new sectors, such as financial services and information systems. However, it has been some time since a large U.S. company attempted to diversify into semiconductor manufacturing, perhaps because of less promising prospects for returns on investment than expected in other fields.

TABLE 2 A Taxonomy of the U.S. and Japanese Semiconductor Industries

United States
Large Systems Houses
  IBM, AT&T/NCR, Rockwell, Hewlett-Packard, Honeywell
  Companies that sell a small percentage of output on the open market but function as captive producers and purchase large quantities of semiconductors from the outside
Systems/Merchant Houses
  Motorola, Texas Instruments
  Companies that sell systems, but whose merchant semiconductor sales account for a high percentage of total sales
Full-Line Houses
  Intel, National Semiconductor, Advanced Micro Devices
Newer Companies
  New Merchants: Cypress, LSI Logic
  Fabless Microprocessor Designers: MIPS, Cyrix
  Fabless Niche Customizers: Altera, Chips and Technologies
  New Systems Emphasis: Sun Microsystems

Japan
Large, Blue-Chip Corporations
  NEC, Toshiba, Hitachi, Mitsubishi Electric, Fujitsu, Matsushita
Other Major Producers
  Canon, Sharp, Sanyo, Oki, Sony
Lateral Entry Companies
  Minebea (NMBS), Kubota, Nippon Steel, Kobe Steel
Smaller Component Makers
  Hosiden, Alps, Sanken, Toko

NOTE: "Fabless" refers to companies which do not possess their own fabrication facilities.
SOURCE: Compiled by NRC Office of Japan Affairs staff.

companies, have been less active in developing strategic alliances with smaller U.S. firms. Whether this is the result of shorter time horizons or a shortage of capital (or both), is open to further study. Certain features of Japanese industrial organization, especially intercorporate shareholding, make it easier for them to operate on the basis of a longer time horizon than U.S. companies. The nature of America's stock market forces U.S. firms to march more closely in step to the drumbeat of quarterly profits.[19] The dense network of intercorporate shareholdings protects Japanese management from the tyranny of short-term profit maximization. Corporate share-

---

[19]Michael L. Dertouzos, Richard K. Lester, Robert Solow and MIT Commission on Industrial Productivity, *Made in America: Regaining the Productive Edge* (Cambridge: MIT Press, 1989), pp. 53-66.

holders, such as banks and insurance companies, do not buy and sell their stocks in response to short-term fluctuations in share prices or to standard indicators such as price/earning ratios.[20] What they seek is the steady appreciation of stock value through sound, long-term company growth.[21]

The high degree of diversification and vertical integration for which Japanese electronics giants are known has also had the effect of heightening interest in strategic alliances because there are potential multiplier effects associated with the introduction of new technologies (through strategic alliances). New technologies can be used to upgrade the quality of both products and process technology for internal, captive markets, including components, manufacturing, software applications, and new product designs. For large Japanese firms, there is a captive market learning curve down which companies can move; the advantages of moving down this learning curve have been widely observed in the case of application-specific integrated circuits (ASICs). Thus, the characteristics of Japanese industrial organization have facilitated Japanese participation in strategic alliances.

BUSINESS CYCLES, TECHNOLOGICAL CHANGE, AND POLITICAL FACTORS

There appears to be a rough correlation between downturns in business cycles and peak periods of alliance formation. When business conditions turn bearish, U.S. companies appear more disposed to enter strategic alliances to meet their financing needs and to survive the sharp downturns in demand without having to revamp existing structures. The alternative is to cut back on R&D projects, new capital investments, and core technical personnel.

Looking at the available data (refer to Figure 3), we see that the number of American-Japanese alliances hit a peak in 1986 and 1987, when business demand in the semiconductor industry slackened. When there was a revival of demand in 1988, the number of alliances plummeted. The empirical evidence is imperfect, of course, making it hard to establish a clear-cut case

---

[20]Masahiko Aoki, *Information, Incentives and Bargaining in the Japanese Economy* (Cambridge, England: Cambridge University Press, 1988), pp. 99-149.

[21]In the early 1990s, a series of Japanese stock market scandals, declining share prices, and deterioration in the positions of commercial banks put an end to the "bubble economy" and rapidly rising asset prices of the late 1980s. Japanese press reports in early 1992 indicated that Japanese electronics firms planned to cut capital investment during the coming year. See "4.9% hen, 6 nen buri mainasu" (4.9% Decline, First Decrease in 6 Years), *Nikkei Sangyo Shimbun*, March 10, 1992, p. 2.

The long-term implications are unclear. Some analysts believe that Japanese companies may be forced to concede large portions of the DRAM business to South Korea and Taiwan. Japanese companies have weathered extended industry downturns in the past.

of cause and effect but the rough, macrolevel data suggest that business cycles do affect the microlevel disposition of U.S. and Japanese companies to take advantage of the benefits of strategic alliances.

A closer look at the data suggests that the situation is more complex than the picture painted by simple time-series analysis. Factors other than fluctuations in business demand can also have significant effects. In 1988 the number of U.S.-Japan alliances fell abruptly from 124 the previous year to 46 (refer to Figure 3). This was the year following the Toshiba Machine Company incident (1987), which involved that company's sale of militarily sensitive technology to the Soviet Union. In the aftermath of the Toshiba incident, MITI officials quietly discouraged Japanese companies from entering strategic alliances or getting involved in overseas activities that would expose them to possible foreign criticism. Hence, political and strategic factors can have a direct impact on alliance formation.

Why in 1989 was there a surge in strategic alliances, when business demand remained brisk? Technological developments leading to the emergence of new product clusters represent another intervening variable affecting alliance formation. The reason strategic alliances increased in 1986-1987 and again in 1989-1990 may have been related to the coming on stream of ASICs, a major new family of products (see Figure 8).

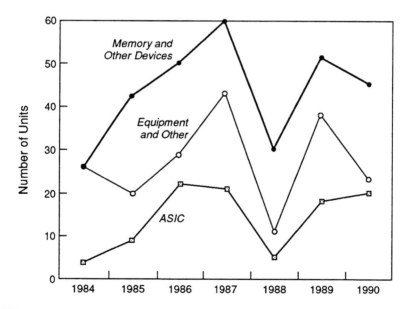

FIGURE 8 American-Japanese semiconductor alliance by product type: 1984-1990. SOURCE: Compiled by NRC Semiconductor Working Group using Dataquest data.

The emergence of ASICs and the passage of an interval of time following the Toshiba Machine Company incident may have provided a stimulus to alliance formation. At any given point in time, a complex mix of underlying and immediate factors comes together to affect trends in strategic alliances.

Mention must be made of yet another major driving force, originating from the fierce competition in the computer industry but pulling semiconductor producers into its vortex: the struggle to define and establish broadly based standards in operating and applications software and the choice of chips (discussed in Chapter 8). This effort to establish operating software standards has driven the recent IBM-Apple alliance aimed against Microsoft. The adoption of MIPS' RISC (reduced instruction set computing) chip and Intel's X86 also provided the framework for the emergence of an ambitious, multifirm consortium known as Advanced Computing Environment (ACE). The struggle over market shares in computers and key semiconductor chips is being shaped by the rush to establish dominance in software and chip standards. This high-stakes race lies behind the alliance alignments and realignments that are helping to redefine the structural boundaries of the semiconductor and computer industries.

RECESSIONS: LESSONS LEARNED

The world semiconductor industry seems to go through a recession roughly every four years. Over the past 14 years there have been four recessions: in 1977, 1981, 1985, and 1990-1991. With each cyclical downturn, semiconductor companies have been forced to make a series of painful adjustments. Accordingly, semiconductor companies have learned some important lessons. Often it takes a serious crisis of some kind to break old, established patterns of doing business, force companies to devise new ideas and approaches, and generate adaptive modes of behavior based on collective learning.

For collective learning to occur, there needs to be some stability in management and in the labor force; the more continuity, the greater is the learning capacity (other things being equal). Japanese corporations may be at an advantage because of their practice of lifetime employment. Owing to the stability of their work force and the continuity of personnel in middle and top management, Japanese corporations are in a position to learn and adapt in ways that elude many U.S. merchant semiconductor houses.

Prior to the 1977 recession, U.S. companies dealt with downturns by cutting variable costs through belt-tightening measures: workers were laid off, operating budgets were cut, and the overall level of activity was curtailed. However, the 1977 recession taught these companies that by itself, variable-cost cutting could no longer be sufficient. The capital intensity of the semiconductor industry had reached high enough levels that mere manipulation of variable costs would not solve the problem. Attention had to be focused on reducing fixed costs as well. This realization prompted companies to attack fixed costs by canceling and slashing plans for expensive new capital investments. The combination of variable-cost cutting plus cutbacks in fixed costs became standard practice from 1977 forward, as evident in the 1981 and 1985 recessions.

Out of the 1977 recession came the further realization that (1) U.S. merchant semiconductor producers share important interests in common; (2) these interests could be effectively served through the organization of an industrial association representing their views in Washington; and (3) although U.S. producers enjoyed technological and commercial dominance (in 1977), the large, diversified Japanese corporations posed a serious, long-term threat. In response to the first two perceptions, the Semiconductor Industry Association (SIA) was organized in 1977; since then, the SIA has shown itself to be among the most active and effective of America's industrial associations in representing member company interests and objectives.

The third perception has led to a sustained focus on Japan as a competitive threat to the U.S. semiconductor industry, a "strategic" industry portrayed as pivotal to America's overall position in high technology.[22] The delineation of this threat has not changed over time; concern over dumping and restricted Japanese market access persists, with heightened concerns about the closed nature of Japanese industrial groupings.

The general tenor of the industry's position toward Japan has not softened since the conclusion of the Semiconductor Trade Agreement in 1986. However, improved relationships have developed between U.S. semiconductor executives and Japanese user executives as direct commercial contacts have been facilitated within the framework of the 1986 agreement and a successor agreement consummated in 1991. The evolution of U.S.-Japan trade negotiations has undoubtedly affected the growth of alliances, although in ways difficult to measure precisely. The 1986 U.S.-Japan semiconductor agreement did not open the door for U.S. semiconductor firms in the Japanese market until the U.S. government applied sanctions on selected Japanese equipment imports in 1987. Although the rate of increase in U.S.-based firms' share of the Japanese market has not been rapid enough to satisfy many observers, the relationships slowly and at times painstakingly nurtured between U.S. and Japanese companies under the agreement have built confidence over time that companies from the two countries can work

---

[22]Semiconductor Industry Association, *The Effect of Government Targeting on World Semiconductor Competition* (Cupertino, Calif.: Semiconductor Industry Association, 1983).

together.[23] In addition, the agreement probably also helped to encourage Japanese firms to form alliances with U.S. firms as a way to ensure their long-term viability in the U.S. market. Although there are many contentious effects of the agreement, such as alleged Japanese dumping in third-country markets in violation of the Agreement, the Agreement stabilized the policy context and thereby probably provided an incentive for alliance formation.

In considering responses to the 1977 recession, it is important to note a different set of adjustments that took place in Japan. Although Japanese firms were subjected to the same underlying cost pressures as their U.S. counterparts, the established practice of lifetime employment precluded layoffs of large segments of their permanent work force (although they did pare down part-time and temporary workers). Japanese corporations were able to continue to invest heavily in new plant facilities, thanks to deep financial pockets (especially the availability of debt financing through the system of main banks).[24] Thus, differences in labor and capital markets led to different corporate responses to recession in the United States and Japan.

The response of Japanese corporations was also affected by another structural difference—the close working relationship between government and industry. In the mid- to late 1970s, MITI's role in organizing national research projects, such as the VLSI project in semiconductors, had the effect of pumping R&D resources into companies at a time of urgent need. This helped them sustain ambitious R&D activities through bad business conditions. Beyond the welcomed infusion of capital, government funding also had the unintended but salutary effect of creating a new, centralized channel of funding for semiconductor R&D, one designed to funnel money centrally from corporate headquarters rather than individual semiconductor divisions. Participation in national research projects also meant that Japanese companies were firmly committed to the completion of these ambitious projects (which could not be terminated because of a public commitment made to the government). It also kept semiconductor R&D elevated at the highest level of priority within Japanese firms. Hence, the impact of government-organized research activities went well beyond the output of the projects themselves. As unintended but important consequences, such national projects left a lasting impact in terms of research commitments and priorities, changes made in the system of corporate R&D funding, and adaptive lessons learned in recession management.

---

[23]Undoubtedly, there is a strong current of opinion in the Japanese semiconductor industry that resents the agreement and sees U.S.-Japan alliances aimed at increasing foreign market share as "gunshot weddings" necessary to avoid U.S. trade sanctions.

[24]M. Therese Flaherty and Hiroyuki Itami, "Finance" in Daniel I. Okimoto, Takuo Sugano, and Franklin Weinstein, eds., *Competitive Edge: The Semiconductor Industry in the U.S. and Japan* (Stanford, Calif.: Stanford University Press, 1984), pp. 134-176.

In the recession of 1981-1982, when the combination of semiconductor revenues and government funding was insufficient to meet their needs, semiconductor divisions in Japanese companies were able to draw on allocations made available by corporate headquarters to underwrite their expenditures. The pool of general funds came from profits earned by the sale of downstream products (consumer electronics products such as video cassette recorders). Semiconductor divisions were thus able to support essential research through cross-subsidizing from profits earned by other divisions. Vertically integrated, diversified Japanese companies, in short, utilized the built-in advantages of cross-subsidization, a natural stabilizing factor during demand downturns in specific product markets.

The biggest and most severe recession to hit the semiconductor industry struck in 1985. The reason this recession hit companies so hard was because business demand had soared to new heights prior to 1985, buoyed by brisk demand from the emerging personal computer market and by abnormally high book-to-bill ratios. The high book-to-bill ratios meant that companies were manufacturing at full capacity and piling up inventory at a fast pace in an effort to meet brisk bookings. When demand dropped abruptly, the feeling of being caught in a free-fall was magnified by the large inventories that companies had amassed (thanks to the false sense of euphoria associated with high book-to-bill ratios). In this sense, one could call the 1985 downturn an inventory-based recession.

Despite harsh business conditions in 1985, Japanese companies maintained very high levels of R&D expenditures. Managers in some Japanese companies took 5 percent cuts in salary and bonuses and curbs were placed on aggressive capital investments. Unlike earlier recessions, there were also cutbacks in plans for the expansion of capital investments. Although Japanese companies went ahead with the construction of such facilities as clean rooms, they chose not to install expensive equipment until the recession had passed. Like their U.S. counterparts, Japanese companies were forced to focus on reducing fixed as well as variable costs.

Japanese companies also discovered that the products embodying older technology tend to be the most vulnerable. Wherever possible, they tried to roll back manufacturing capacity in products based on slightly older technology and to push aggressively ahead in products incorporating newer technology. By moving from the rear to the forefront, Japanese companies anticipated gaining some measure of insulation against cyclical free-falls. American firms already understood that. As a result, companies in both countries moved to close down excess capacity in older products during the most recent recession (1990-1991). To make this move, attention turned again to the advantages of joining forces with well-chosen corporate partners.

Mired in a deep recession, U.S. companies had no choice but to make deep cuts in both variable and fixed costs, laying off workers in record numbers and canceling plans for large-scale capital investments. American and Japanese firms began to see strategic alliances as a practical, countercyclical option—as either an alternative or a supplement to other cost- and risk-reducing measures. Many U.S. firms came to see strategic alliances as a means of obviating the need to make certain manufacturing, marketing, or R&D investments that they might otherwise have felt compelled to make.

Around the mid-1980s when the number of strategic alliances soared, the impetus grew to pool resources to fuse one company's strengths with those of a partner so as to consolidate and expand competitive advantages while offsetting company weaknesses. The Toshiba-Motorola alliance is a typical example explored in more detail in the case study that appears in Appendix A.

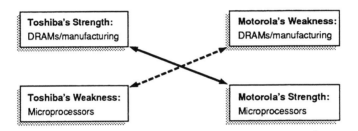

By bringing together complementary strengths, the partners filled out their product portfolios, enhanced their overall technological capabilities, and compensated for in-house gaps and weaknesses. The scope of strategic alliances broadened and extended into virtually all facets of the semiconductor business:

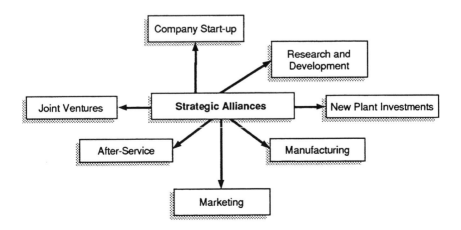

Once the advantages became apparent, no phase of the semiconductor business could stand outside the powerful pull of alliance linkages. Forming alliances became a fixture in the range of corporate options:

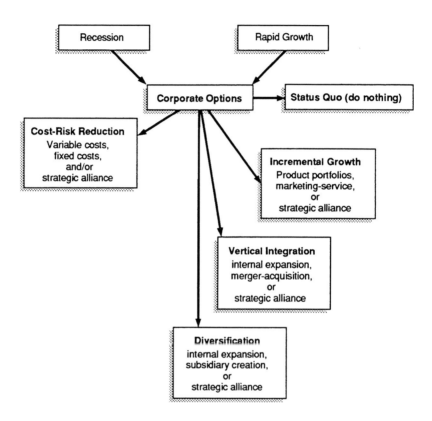

Increasingly, the advantages of alliance formation came to outweigh the perceived risks and uncertainties. Indeed, the costs of going it alone (i.e., standing aloof from alliances) came to be seen as unacceptable. To survive in the competitive marketplace, especially during recessions, semiconductor companies had to scramble to find strategic allies. The rush to find allies, in turn, further intensified the level of commercial competition.

5

A Typology of Alliances

If the 16 alliance types described in Chapter 3 are listed by frequency of occurrence in 1990, they can be rank-ordered as illustrated in Figure 9. The order is not surprising in terms of where the U.S. and Japanese semiconductor industries stand today. One would expect fabrication contracts, sales/marketing agreements, and joint development projects to be among the most actively sought and commonplace forms of bilateral alliances, given the current commercial situation. Still, the relatively small number of technology exchanges and patent licensing agreements is somewhat surprising, because these offer the simplest and most direct routes to technology transmission. For several decades, licensing of technology through sales of rights to patents was the sole form of U.S.-Japan semiconductor alliance, and the evolution of new and more complex alliances reflects the maturation and complexity of the industry itself.

To understand the impact of various types of strategic alliances on technology transfer, we can modify and expand our original classification, using two criteria on which to construct a typology: (1) the level of commitment to repeat transactions over time (taking into account the ease or difficulty of exiting from various types of strategic alliances) and (2) the degree of closeness or joint involvement and common organizational membership (organizational fusion). The various types are listed in Table 3.

The first cluster in Table 3 is the easiest to enter and exit, and constitutes the largest category of U.S.-Japan alliances (see Figure 10) but technology transfer is limited largely to one-shot transactions. The opportuni-

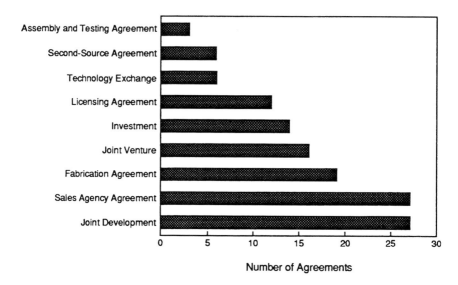

FIGURE 9 American-Japanese semiconductor alliances by agreement type: 1990. SOURCE: Bruce Kogut and Dong-Jae Kim, "Strategic Alliances of Semiconductor Firms," unpublished report to Dataquest, January 1991.

ties for more sustained technology transfer are built into the second cluster of alliances, repeated transactions, but this represents the smallest category. The optimal framework for continuous transfer is embodied in the third cluster, the organizational fusion category, featuring some form of common organizational membership. Despite high entry barriers, a surprising number of alliances fall into this category. It should be pointed out, however, that alliances in the third cluster require the highest levels of up-front investment and risks, and are the hardest to organize and manage (but once organized, the most apt to be sustained). Still, movement toward the third cluster increases the likelihood of long-term interaction and a two-way flow of technology.

What separates cluster I alliances from those in clusters II and III is that the latter two give rise to quasi-formal or formal structures of continuous technology transmission over time. The alliance types in cluster II (especially joint development projects) offer limited opportunities for continuing or two-way technology transmission. Because of the seminal and growing importance of software, the formation of strategic alliances based on standards coordination is becoming increasingly common and important in terms of giving shape to the rapidly changing marketplace. The third cluster of alliance types, including joint ventures, product/technology-seeking investments, mergers, and acquisitions, encompasses the most enduring

TABLE 3 A Classification of Strategic Alliances Types

I. Free Exit/Arms Distance
Purely profit-seeking investments
Sales and marketing arrangements
Procurement

Assembly and testing
Second sourcing
Fabrication agreement

Contract research
Consulting services
Patent acquisition
Technology exchange
Cross-licensing

II. Repeated Transaction/Quasi-Formal Bonds
Standards coordination
Regular exchange of technical personnel
Joint development

III. Costly Exit/Organizational Fusion
Joint ventures
Participation in research consortia
Product/technology-seeking investment
Mergers/acquisitions

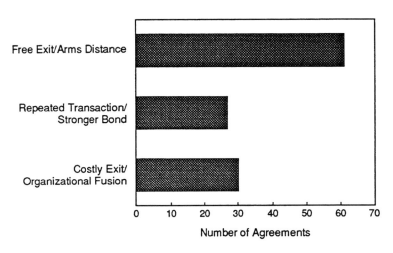

FIGURE 10 Alliances by cluster grouping: 1990. SOURCE: Compiled by NRC Semiconductor Working Group using Dataquest data.

organizational forms for ongoing technology transfer (even though the scope and flow of transfer depend on a variety of exogenous factors).

As the semiconductor industry matures, more alliances of the cluster III type are forming. From the standpoint of alliance stability and the potential for greater mutuality of benefits, the direction is encouraging. Whether it actually leads to a greater inflow of technology from Japan that enhances America's technological infrastructure remains to be seen. What the U.S. semiconductor industry as a whole needs (a stronger manufacturing infrastructure) may not be transmitted effectively through the mechanism of private company tie-ups. Alliance agreements may fulfill individual company needs but not meet collective needs. Here again, there may be a disconnection among corporate objectives, industry-wide needs, and national interests. It may take an industrial association (such as the SIA), a business federation, or the government to call attention to the broader, longer-term needs of the semiconductor industry as a whole and to encourage companies to structure alliances to ensure that the country's infrastructure is strengthened.

In Japan industrial associations and MITI play that kind of role. They remind private companies of broad, collective interests and goals in seeking to harmonize activities at the three levels of company-industry-nation. The fate of a key industry such as semiconductors is far too important to leave completely to the market mechanism (no matter what laissez-faire economists say). In the United States too, there may be a need for "extramarket" consensus building, harmonization, and encouragement to contribute to the collective good, although the mechanisms will necessarily be far less formal than in Japan.

DIRECTION OF TECHNOLOGY FLOW: NATIONAL INTERESTS

Outside the realm of formal alliances, a variety of informal channels exist through which technology transfer occurs. Foreign companies seeking to obtain technology have direct and open access to knowledge in the public domain—university-based research, university-industry research consortia, graduate training, contract research, academic consultation, scholarly and business journals, professional conferences, patent application disclosures, personal relationships, and so forth. Although measuring the volume of one-way technology outflow through such public channels is impossible, the fragmentary and unsystematic evidence (such as the number of foreign nationals receiving graduate training in the United States) suggests that the volume of flow is heavy—perhaps heavier than the one-way transfer through strategic alliances.

Contract research and consultants commissioned by Japanese companies also have the effect of moving knowledge into Japanese hands (unfortunately, data for such transactions are unavailable, and there is no way of knowing how widespread these practices have been). Second sourcing and fabrication agreements have served to transfer knowledge from small U.S. start-up firms to large Japanese semiconductor manufacturers. Similarly, although to a much lesser extent, the same can be said for assembly and testing agreements.

At bottom, the decisive factor driving such one-way transfers may be comparative weaknesses in U.S. manufacturing infrastructure and processing technology. The Japanese and Asian edge in manufacturing across so many different sectors may mean that product niche technologies will continue to move out of the United States into Japan and Asia. Industry leaders note that U.S. industry has greatly increased its emphasis on quality and manufacturing productivity in the past few years.[25] In 1990, cluster I alliance types accounted for 35 percent of all bilateral activities in semiconductors.

Cluster III also encompasses alliance forms that tend to push technology in one direction. Acquisitions, mergers, and equity investments have the explicit objective of obtaining new products and technology. The abundance of capital in giant Japanese companies, their willingness to invest in the United States, the small U.S. start-up companies' desperate need for capital to survive, the striking differences in industrial structure, and the system of permanent employment (which helps to insulate Japanese companies from outside take overs) provide incentives for such Japanese investment. Over time, these factors may also erode the capacity of U.S. firms to invest as extensively in Japan as the Japanese invest in the United States. Japanese corporations have accounted for nearly all of the equity-related transactions in semiconductors over the past five years.

By their very nature, certain types of strategic alliances generate a two-way flow of technology. Included here are technology swaps, cross-licensing agreements, personnel exchanges, joint development projects, and joint ventures. Such positive-sum alliances grow out of a commonality or convergence of corporate interests and goals. Technology swaps, cross-licensing (cluster I), and regular personnel exchanges (cluster II) are common and relatively simple alliances to conclude; the costs and risks tend to be low but so too, by the same logic, are the advantages and benefits. Joint development projects and joint ventures represent far more challenging and complicated types of strategic alliances. They usually grow out of a history of mutually satisfactory interactions or the powerful pull of a common, overriding threat or a common customer base.

---

[25]Some leaders believe that the SEMATECH consortium, with 14 U.S. member companies sharing the $200 million annual cost with the U.S. government, has been a key factor in helping to narrow the quality and manufacturing gap with Asian producers. Demonstrable evidence will come as U.S. manufacturers in larger numbers offer competitive manufacturing and second-sourcing services for U.S. firms.

The number of joint ventures is still relatively small. In 1990, joint ventures represented only 14 percent of all U.S.-Japan strategic alliances. In view of the formidable difficulties of organizing and sustaining successful joint ventures (such as deep-seated differences in corporate cultures), the paucity of joint ventures is hardly surprising. Indeed, considering the very small number of joint ventures that have managed to survive the passage of time, we can understand why more joint ventures fail to get organized. The mere fact that a strategic alliance takes the form of a joint venture does not necessarily mean that technology will be exchanged; sometimes the direction of flow is lopsided. Nevertheless, in those few cases where joint ventures have succeeded, the mutual benefits to both sides tend to be substantial.

Joint development is another form of strategic alliance that can have the positive-sum effect of enlarging the overall pie. If organized properly, joint projects function to diffuse technology in two or more directions. Like joint ventures, joint development alliances tend to grow out of a history of mutually satisfactory transactions (e.g., iterative procurements and second sourcing). Most joint development projects in 1990 were based on central processing units (CPU) and microprocessing units (MPU), with relatively little emphasis on application-specific integrated circuits (ASIC) and memory chips (see Figure 11). By contrast, memories and ASICs dominated manufacturing alliances. Thus, by their nature, joint development alliances

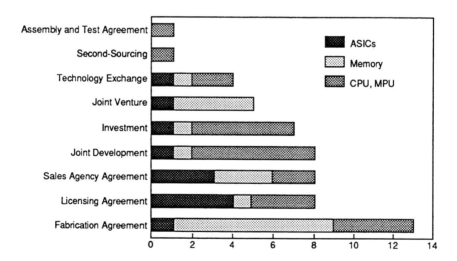

FIGURE 11 Agreement type and product technology: 1990. SOURCE: Bruce Kogut and Dong-Jae Kim, "Strategic Alliances of Semiconductor Firms," unpublished report to Dataquest, January 1991.

tend to focus on the more complex technologies of the future, whereas manufacturing alliances—second sourcing, fabrication, assembly, and testing—concentrate more on commodity chips.

What all this implies is that strategic alliances are not random occurrences. Cluster I types are the simplest alliances to enter into; these are the alliances that companies with no previous contact are most inclined to conclude. The two-way types of alliance in clusters II and III tend to follow a linear pattern, building on the foundations laid by cluster I agreements. There is a logical, step-by-step progression, as shown in Table 4.

The scope of interaction, mix of risks and rewards, and level of commitment and organizational complexity increase as companies move forward from one phase to the next. Leap frogging sometimes takes place, especially when small U.S. start-up firms (with no previous experience with Japanese companies) are involved but the modal pattern is one of steady progression based on an unfolding process of interaction and mutual trust. At any time, the progression can be halted or reversed but the costs of exit increase as one advances along the continuum.

The progression from joint development to joint venturing is a monotonic jump, one that sharply ratchets upward the costs and risks of joint action (as well as the potential benefits). For this reason, joint development represents a kind of natural "resting" or stopping point along the alliance continuum. Relatively few companies cross this threshold. Because joint development projects can be finely targeted and do not require permanent union but can be disbanded, the costs, risks, and difficulties tend to be significantly lower than they are for joint ventures. American and Japanese companies, bringing complementary technological and marketing strengths to the table, frequently find that they have strong incentives to launch joint development projects, particularly if they share a common customer base or operate on the same technical standards.

In 1990, joint development projects represented the most common form of strategic alliance, accounting for nearly one-quarter of the total. Without longitudinal data it is impossible to discern trends, but this type of alliance

TABLE 4 Alliance Progression

Early Contact	Regular Transactions	Common Endeavors	Joint Ventures
Patent licensing Cross-licensing Marketing/ distribution	Procurements Fabrication Second sourcing	Joint development	

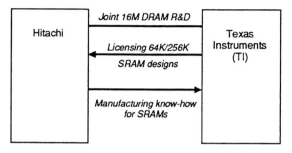

In 1989 Hitachi and TI announced they would expand their alliance in which
1. They would jointly develop 16M DRAMs,
2. Hitachi would license TI's 64K and 256K SRAM designs and provide manufacturing technology for the chips, and
3. Both companies would relabel each other's chips for sales.

Trade-offs for TI	
Benefits	Costs
R&D cost savings	Sacrifice older SRAM designs
Access to Hitachi's process know-how	Potential loss of sales to Japanese rivals
Credibility of Hitachi as a partner	

FIGURE 12 Joint R&D—Texas Instruments and Hitachi. NOTE: In 1991 Hitachi and TI announced an intention to extend the alliance to 64M DRAMs using a common design. SOURCE: Sheridan Tatsuno, NeoConcepts, 1990.

appears to be increasing steadily over time and is likely to continue to rise. The calculus of costs, risks, and benefits can be very favorable, relative to the other forms of strategic alliances, if companies bring complementary strengths to the table. The Texas Instruments-Hitachi alliance (1989) is a good example (see Figure 12).

The impact of certain forms of strategic alliances on the one- or two-way flow of technology is shown in Table 5. Whether, and to what extent, the various alliance mechanisms facilitate one-way or two-way technology transfer depends, of course, on the specific provisions of the agreements. In certain cases, joint ventures will promote two-way technology transfer; in other cases, technology will be captured by only one side.[26] The mechanism itself is sometimes not the decisive factor, but various types of strategic

---

[26] Robert B. Reich and Eric D. Mankin, "Joint Ventures with Japan Give Away Our Future," *Harvard Business Review*, March-April 1986, pp. 78-85.

TABLE 5 Alliance Impact on Technology Flow

One-Way Flow	Neutral	Two-Way Flow
Patent licensing	Strictly financial investment	Technology exchange
Assembly/testing		Cross-licensing
Second sourcing	Sales/marketing	Personnel exchange
Fabrication	Procurement	Joint ventures
Contract research		Research consortium
Consulting services		Joint development
Standards coordination		
Product/technology-seeking equity investment		
Merger/acquisition		

alliances have built-in propensities. What ultimately determines the direction of technology transfer are the specific terms upon which alliance partners agree.

ASYMMETRICAL PAIRINGS: LARGE AND SMALL COMPANIES

By disaggregating the 1990 data on strategic alliances according to size of companies, it becomes clear that the vast majority of U.S.-Japan alliances are between small- or medium-sized U.S. companies (many of them young start-ups) and large, vertically integrated, diversified Japanese corporations. Alliances have been struck between large U.S. and large Japanese corporations, but the number is much smaller than those involving small and large pairings (see Figure 13). A few alliances have been concluded between small U.S. and small Japanese companies, but these are of scant significance from the standpoint of technology transfers. Finally, an increasing number of alliances involve large U.S. firms such as IBM, DEC, AT&T, and Hewlett-Packard linking up with small U.S. firms.

It is the alliance of small U.S. firms and large Japanese corporations, the most prevalent form of U.S.-Japan linkage since 1980, that is thought to transfer the bulk of American know-how to Japan. Looking at the question from the microlevel perspective of small U.S. companies, we can glean reasons for what is sometimes myopic and self-defeating behavior on the part of these firms.

Perhaps the primary explanation for the propensity of U.S. companies to "give away" technology lies in the glaring asymmetry in the size and staying power of small U.S. firms, on the one hand, and of large, diversified Japanese *kaisha*, on the other. In negotiating deals with Japanese giants,

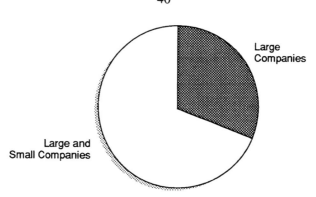

FIGURE 13 Alliances by company size: 1990. SOURCE: Compiled by NRC Semiconductor Working Group using Dataquest data.

small U.S. firms usually bargain from a position of overwhelming weakness, given their fragility and lack of independence. To get what they need, they have often felt that they had to give up their only valuable asset—marketing rights for new niche products or leading-edge know-how (especially in software). If they were unwilling to share their know-how, potential Japanese partners would not have the slightest interest in striking a deal.

By leveraging technology, small U.S. start-up firms sometimes try to survive from month to month. On the list of most urgent needs are the following: obtaining large infusions of capital; finding outlets to market, distribute, and service their products; and advancing their R&D activities to compete in the next generation of products. For small U.S. firms, efforts to strike a balance between conserving negotiable assets and meeting immediate needs have often been futile. Simply staying afloat requires that they bend all their resources and energies to meet their urgent needs (see Table 6).

Even if a small company happens to be cash rich or have access to low-cost capital—which is definitely not the typical pattern—and even if it possesses a "hot" product or seminal new technology that attracts eager suitors, the need to come up with the best manufacturing arrangement and to find marketing, distribution, and servicing networks in Japan's challenging market often makes it imperative that small U.S. firms connect with one or more large Japanese corporate partners. The up-front costs, downside risks, and enormous difficulties associated with independent attempts to break into the lucrative Japanese market are simply too daunting to ignore.

In many cases, the bargaining power of small U.S. firms is also limited by such factors as (1) the press of time and a strong sense of urgency occasioned by short-term time horizons; (2) a lack of experience and knowledge in dealing with Japanese companies; (3) the difficulty of playing one potential Japanese partner off against others; (4) the reactive posture of most small U.S. companies, which typically wait passively for Japanese companies to approach them; (5) the propensity of small U.S. firms to lock themselves into alliances with the first Japanese suitors to make a proposal (without checking out other potential partners); and (6) the difficulties of breaking into established networks in Japan.

Although many of these disadvantages can be counteracted if the small firm has good advice from those familiar with other U.S.-Japan alliances, including provisions for dealing with intellectual property rights (IPR), many small U.S. firms have not received such advice. American companies (large and small alike) often enter into alliances with only limited, short-term goals in mind. Nor have small U.S. firms routinely entered alliances with the conscious intent of leveraging what they learn from the Japanese partner to enhance and consolidate their strategic position down the road. By failing to go through a process of organizational learning, U.S. firms incur major opportunity costs and eventually find that they have not benefited as much as their Japanese partners. Their shortcomings stem from both structure and strategy.

In stark contrast to small U.S. start-up firms, established *kaisha* possess an impressive array of negotiable assets, their needs are not nearly as pressing, and their whole approach to foreign alliances is long term and strategic. To *kaisha,* strategic alliances represent an indispensable tool for gaining immediate commercial advantage and moving down a longer-term learning curve. Alliances are not mandatory in the sense that their short-run survival depends on them, as is often the case for small U.S. start-up companies. In this respect, the *kaisha*'s balance sheet of assets versus needs is often the mirror image of that for small U.S. start-ups: assets far outweigh liabilities and needs (see Table 7).

TABLE 6 Characteristics of Small American Companies

Assets	Limited and time bound
	Leading-edge or niche product and/or technology
	A few high-powered researchers
Needs	Urgent and multiple
	Large amounts of capital
	Timely evidence of profitability
	(or prospect of future profitability)
	Manufacturing capabilities or access
	Marketing/distribution/service (especially overseas)
	Continual development of next-generation products
Time horizon	Short term and not bent on organizational learning

TABLE 7 Characteristics of Large Japanese Corporations

Assets	Deep and multifaceted
	Abundant capital
	Strong bank backing
	Intercorporate shareholding pattern
	Relative insulation against pressures for short-term profitability
	Manufacturing excellence (costs/quality/flexibility/turnaround time)
	Strengths in process technology
	Extensive networks for marketing, distribution, and servicing (especially in Japan and Asia)
	Vertical integration
	Horizontal diversification
	Strong subcontracting networks
	Keiretsu organizations
	Collectivization of risks, costs, and profits
	Power of demand-pull procurements
	Large patent holdings
	Close working relations with bureaucracies
Needs	Contained, not desperate
	Promising new products/technologies
	Coverage for in-house gaps/deficiencies
	Perception as a good international corporate citizen, not a predatory trading partner
Time horizon	Long term and bent on organizational learning

As latecomers, the *kaisha* have always shown a remarkable receptivity to outside ideas, but the reasons go deeper—to structural factors having to do with disparities in levels of vertical integration and scope of diversification, as well as to differences in management and corporate organization. Japanese companies are larger, more vertically integrated and diversified, have more financial resources, and are more aggressive in their pursuit of alliance opportunities abroad than U.S. firms. They also have a wealth of experience in dealing with small- to medium-sized companies, because many *kaisha* have extensive subcontracting networks with small- and medium-sized Japanese companies. Some of this experience of asymmetric interaction is relevant to their relationships with small U.S. firms.

Almost as much as structural factors, however, the attitude and intent of Japanese companies go a long way toward explaining imbalances in the distribution of benefits. Japanese firms view strategic alliances from a longer-term time horizon and enter the relationship with the intention of learning as much as possible in order to strengthen their competitive position several years down the road. Their capacity for organizational learning is matched by only a small handful of American companies. Whereas U.S. firms may gain access to low-cost fabrication facilities or realize short-term increases in sales from strategic alliances, their Japanese partners often have bigger objectives in mind, such as applying what they have learned to the development of whole families of new products.

Many of the *kaisha* have established within their organizations new product divisions, which have no corresponding units in U.S. companies. These new product divisions have the responsibility of ferreting out all possible opportunities to add products and technologies to the company's extant portfolio. Not only do these divisions assess concrete opportunities in-house, looking at the range of commercial applications of technology in their company's R&D pipeline, they also thoroughly explore and aggressively pursue all commercial opportunities overseas, including those offered by small U.S. start-ups. It is, in part, because of such organizational differences (reflecting broader differences in corporate strategy) that small U.S. companies wind up wedded to Japanese corporations.

By comparing the assets and needs of small U.S. firms and large Japanese corporations, we can understand why fledgling U.S. companies find themselves at a marked disadvantage in negotiating deals with *kaisha*. There are signs, however, that small U.S. firms are beginning to develop more savvy approaches to structuring their alliances with large Japanese *kaisha*.

Small U.S. firms developing technologies that have application to products designed in the United States or, at least, are controlled in the market by U.S. standards, such as operating systems standards, networking standards, interface standards, and communications standards, may be in a relatively good position to build alliances with Japanese companies. Such U.S. semiconductor companies tend to focus on chip design. They have the skill required to design a competitive set of chips that serves a focused customer need, and they can protect their intellectual property with strong patents. These companies do not push the frontiers of process technology; they use what is available from foundry vendors. Until recently, most of these foundry relationships have been with Japanese companies with large fabrication ("fab") plants. Some have asserted that "fabless" companies have not contributed to the strength of the U.S. industry, but they have added value at one stage of the strata. Further, their success in the market has encouraged several U.S.-based wafer suppliers (such as AT&T and Philips/Signetics) to open their fabrication facilities to these firms, leading to increased U.S. content.

Another group of U.S. semiconductor companies include those that make a mark by pushing forward the state of process art. Cypress Semiconductor, which used its initial public offering to build its own state-of-the-art fabrication plants, successfully utilized this strategy. In the last five years there have been examples of fabless companies whose raison d'être has been improvement in some process area.

One example is Power Integrations of Mountain View, California. The founders figured out a way to make very high-voltage transistors that consumed very little chip area by using the same processes commonly used to manufacture normal, low-voltage integrated circuits. Power Integrations has developed a way to integrate power supply and logic. The market for power supplies is very large because every electronic system from a complex mainframe computer to an electric toothbrush needs a power supply. This technology is also useful in producing "smart" battery charger products—an important requirement for the increasing number of battery-powered electronic computing and communications products.

Power Integrations aimed to build a large semiconductor company on the venture capital it raised. It needed a large supply of wafers, and success would depend on its ability to control its proprietary technology. First, Power Integrations proved its process by building devices at Orbit Semiconductor, a small-run fabrication house in Silicon Valley that produces wafers on a quick-turn basis for fabless semiconductor firms. Next, it shopped for a large fab partner. Realizing that its market was global, it did not limit its search to the United States and did not want to grant a broad license to its technology. It found Matsushita of Japan. Matsushita would be a large customer for its products in the consumer electronics and computer divisions. A deal was struck in which Matsushita will build wafers for Power Integrations at a competitive price and will be licensed to use the technology internally, for which it will pay Power Integrations a fee and royalty. Power Integrations will maintain the right to sell products to other Japanese customers. An alliance has been formed. Power Integrations gets capital in the form of a license fee, loses the market of one large potential customer, and maintains the rights to its technology and patents. The company has since entered into a similar arrangement with AT&T. The need for volume supply drove Power Integrations into this type of deal, whereas the strength of its patents and the present commitment of the courts to respect IPR made larger companies willing to enter agreements that limited their use of the technology and authenticated Power Integrations' position by paying a substantial fee to it.

TriQuint Semiconductor, a gallium arsenide (GaAs) semiconductor company based in Beaverton, Oregon, is another example of a small U.S. company developing technology-based strategies for global marketing. The company was formed through the merger of the TriQuint subsidiary of Tektronix and two venture backed companies—Gazelle Microcircuits and Gigabit Logic. The GaAs business is small now, but its potential is large because of the growing interest in high-performance computing and mobile communications. The Japanese have been active in GaAs for many years because it is an important technology for some display applications. In the United States, the early backing for GaAs came from the Defense Advanced Research Projects Agency (DARPA), which was interested in the communications capability of the technology. For TriQuint, Japan is a natural market of interest, although all the major Japanese computer and communications companies have their own GaAs efforts. In order to agree to acquire the products, potential partners have insisted on an extensive license, so no alliances have been formed to date. Even though TriQuint is in need of capital, it has chosen to regard the price as being too high if technology is to be granted in exchange for capital. American interest in the field is growing, and another GaAs company, Vitesse Semiconductor, recently completed a successful public offering to meet its capital needs. Although 15 years ago, U.S. companies would have given up the Japanese market and their technology to get Japanese capital, that is not the case today.

A final example is nCHIP, a company engaged in the new field of high-performance multichip modules. The founders were originally associated with Lawrence Livermore National Laboratory and the company has a technology with good patent protection for building silicon subsystems that operate at more than 100 MHz. nCHIP's substrate for the packaging technology is silicon, and its success depends on the low-cost supply of a large volume of silicon substrates. Although users are excited, it has been difficult to fund this effort because high-volume demand is still several years away. The company went to Japan in search of an alliance that would solve its supply and capital requirements, and structured an agreement with Sumitomo Metal and Mining (SMM) in which SMM is granted internal use of the technology for its subsystem products and makes an equity investment in nCHIP. The license in this case is more extensive than in the case of Power Integrations because the capital needs are greater and thus more had to be bartered. "Pure" equity with a longer time window would have been preferable, but the market will be large and the agreement maintains nCHIP's rights to market directly in Japan.

Capital is available in the United States to start new semiconductor firms if their contribution is considered focused on a specific issue in the strata of semiconductor technologies. Since capital is not available to build fully integrated firms, alliances have become a way of life. Large U.S. companies increasingly recognize that alliances with start-up firms may be beneficial to their full product lines, and they are reviewing proposals made by these start-ups. Meanwhile, Japanese venture investments are not increasing at the pace they did a few years ago.

Small U.S. firms must continue to pursue markets in Japan because of their size. These examples illustrate that small U.S. semiconductor companies may be moving up the learning curve in their interactions with Japan. Companies based in other Asian countries are increasing their efforts to enter into alliances with small U.S. companies. These overtures are viewed favorably because of the significant semiconductor capability that now exists in Korea, Taiwan, and Singapore, countries that are presently active in seeking alliances. The growing array of players in the global semiconductor industry thus offers alternatives to U.S. companies.

SYMMETRICAL PAIRINGS: LARGE COMPANIES

If size disparities place small U.S. start-up firms at a disadvantage, the playing field is more level when large American and Japanese companies enter into alliances. Most of the large-large alliances involving electronics companies (as opposed to companies trying to diversify into electronics) have served to transfer technology in two directions. Often, U.S. companies ask Japanese companies to second source, fabricate, sell, or jointly develop leading-edge products, such as RISC chips and microprocessing units, as in the case of the Mitsubishi-AT&T alliance. Another common form is a technology swap, with Japanese companies supplying know-how in dynamic random-access memory (DRAM) or static random-access memory (SRAM) in exchange for American know-how in MPUs or ASICs (as in the case of the Toshiba-Motorola alliance). In most cases, both sides come to the table with clear bargaining assets, looking to conclude agreements that will fill in niches, lighten cost burdens, and spread out risks.

In contrast to small start-up firms, the large U.S. electronics companies do not have their backs against a wall and are capable of assimilating new products or technologies from the outside. Yet even when large U.S. corporations are involved, fundamental deficiencies in America's industrial base—especially weaknesses in its manufacturing infrastructure, the short-run imperatives of U.S. capital markets, the semiconductor industry's overreliance on the computer industry, and the absence of a substantial presence in consumer electronics—lead to difficulty in insuring that they make the most of alliances with Japanese companies. According to Robert Reich, U.S.-Japan strategic alliances, even those bringing together partners of roughly equal bargaining power, have resulted in transfers of state-of-the-art technology to Japanese competitors.[27]

Not many alliances have been consummated between large Japanese companies outside the electronics industry and established U.S. electronics companies. Large Japanese companies such as Kubota and Nippon Steel seeking to diversify into semiconductors invest in, acquire, or hook up with small U.S. firms. Kawasaki Steel's joint venture with LSI Logic illustrates the balance sheet of benefits and costs that accompany diversification through

---

[27]Reich and Mankin, op. cit.

organizational fusion (see Figure 14). These alliances represent a substantial share of the total number of U.S.-Japan semiconductor alliances.

From the standpoint of Japanese corporations, it is generally easier to deal with smaller U.S. firms. Provisions to acquire and transplant semiconductor technology are more easily arranged. Japanese companies can get more benefit from their investments, stringing together a series of bilateral deals involving several small U.S. companies offering technology in different areas. Having the flexibility to pick and choose partners based on specialized products and technologies—what might be called "boutique technology shopping"—can be a major advantage compared to being locked into substantial deals with large, established U.S. corporations. Dealing with small U.S. firms entails certain risks, in light of their sometimes precarious existence, but most Japanese investors conclude that the advantages outweigh

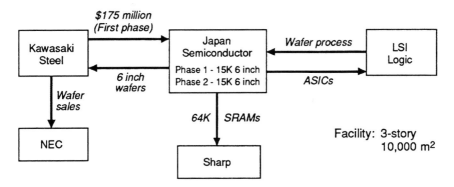

Japan Semiconductor (NSI) is a joint venture established in 1985 by LSI Logic and Kawasaki Steel, whose first manufacturing facility in Tsukuba began operations in 1987. Kawaski Steel made an initial investment of $175 million and owns 45% of NSI, while LSI Logic transferred its wafer process from its California plant and holds 55% equity. NSI supplies ASICs and SRAMs to its parent companies and to outside vendors. NSI's $155 million second phase facility was built last year but the start of mass production was postponed until March 1993 because of the industry recession. Orders are expected to pick up, and NSI aims to reach $150 million in sales in fiscal 1993.

Trade-offs for LSI Logic

Benefits	Costs
Solid Japanese partner	Differing corporate cultures
Kawasaki long-term investment	Different product goals
Kawasaki not a major integrated circuit rival	Long start-up cycle

FIGURE 14 Joint venture—Kawasaki Steel and LSI Logic. SOURCE: Sheridan Tatsuno, NeoConcepts, 1992.

the risks. Perhaps the clearest advantage is that Japanese corporations have more say in dealing with management in small U.S. firms.

Only in exceptional cases, such as the alliance between Kobe Steel and Texas Instruments (involving a joint venture in ASICs), are large, nonelectronics Japanese firms willing to tie up with large American semiconductor companies. When such large-large tie-ups do occur, they tend to take the form of joint ventures or joint development projects—cluster II and III types of strategic alliances. This means that they give rise to relatively complicated organizations requiring Japanese companies to bring some technological capability to the joint undertaking.

Japanese companies wishing to cross the sectoral divide and diversify into semiconductors have a hard time doing so from core businesses that are technologically unrelated. Japanese companies in what might be called the assembly and processing sector—steel production, automobiles, machinery, and precision equipment—possess the kind of technological and manufacturing infrastructure that can be used to launch into semiconductors, whereas those in sectors such as basic materials—oil, energy, textiles, agriculture, cement, paper, and pulp—do not have the requisite infrastructure for diversification. Nearly all cases of cross-sector movement into semiconductors have involved Japanese companies from the assembly-processing sector.

WHY DIVERSIFICATION?

To date, Japan is one of the few countries in the advanced industrial world in which large, established companies in smokestack sectors (steel), old-line manufacturing (machinery, automobiles), and skilled assembly (precision equipment) have made the difficult transition into the high-tech world of semiconductors and electronics (see Figure 15). Among the Japanese companies that have already made the transition are Nippon Steel, Kawasaki Steel, Kobe Steel, NKK (Nippon Kokan), Sumitomo Metals, Kubota, Minebea, Toppan Printing, Toyota Motor, Hoya Glass, Yamaha, Cannon, and Seiko. Comparable examples of lateral, cross-sector mobility in the United States or Europe are hard to find.

Horizontal diversification is a vehicle for technology transfer from the United States to Japan; within Japan, diversification serves as a mechanism for scattering technology widely across industrial sectors. Cross-sectoral diffusion is particularly important in terms of generating new commercial opportunities. Traditionally, the Japanese company has been seen as reluctant to move away from core businesses; when diversification occurred, it was through the creation of semiautonomous subsidiaries.[28] Seiko, for

---

[28]Rodney Clark, *The Japanese Company* (New Haven: Yale University Press, 1979), pp. 53-64. See also Iwao Nakatani, *The Japanese Firm in Transition* (Tokyo: Asian Productivity Organization, 1988), pp. 87-93.

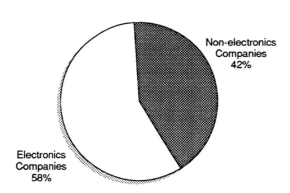

FIGURE 15 Investments by type of company 1986-1990. SOURCE: Compiled by NRC Semiconductor Working Group using Dataquest data.

example, diversified into semiconductors and computers through the creation of a subsidiary, Epson.

What is it about Japan that accounts for the interest in recent years in diversification? Perhaps the most obvious reason is the Japanese practice of lifetime employment. When it becomes clear that their core businesses will lose comparative advantage, there are built-in and compelling reasons for the *kaisha* to begin branching out into promising new fields of business activity. Diversification enables them to retain their work force, utilize sunk investments, and survive as corporate entities. Stated in causal terms, the loss of comparative advantage is the catalyst (the necessary condition), and the distinctive features of Japanese industrial organization give shape to the company's response (the sufficient condition).

Some Japanese corporations may try to delay the inevitable as long as possible through government intervention (such as trade protection and subsidization). However, since MITI is disinclined to protect declining or sunset industries for extended periods, Japanese companies have little choice but to explore opportunities for survival through diversification. The combination of lifetime employment and intercorporate stockholding has the effect of closing off alternative routes of survival, such as mergers and acquisitions involving other Japanese companies.

The underlying forces at work pushing Japanese firms to diversify should be kept analytically distinct from the factors that account for success or failure in implementation. Examining how and why Japanese steel companies have succeeded in producing high-quality semiconductor chips is a fascinating question, since it would be hard to imagine U.S. Steel or British Steel demonstrating the same capacity to turn out DRAMs or ASICs, but the subject is beyond the scope of this report.

The need to diversify has been one of the driving forces behind a noteworthy trend in U.S.-Japan strategic alliances: namely, the proliferation of tie-ups between large Japanese corporations outside of electronics and both small and large American electronics companies. Whether these alliances will be sustained during a period of downturn and whether they will result in the building of substantial technical capabilities on the part of the large Japanese corporate partners are key questions for the future.

6

Issues for U.S. Policy: Japanese Investments and U.S. Competitiveness

Dataquest statistics indicate that investments in 1990 accounted for only 12 percent of the U.S.-Japan strategic alliances. The American Electronics Association (AEA), however, has identified a much larger number of such alliances over a five-year period—500 cases of Japanese investments in America's electronics industry between 1986 and early 1992.[29] The AEA listing provides support for those who say that such investments constitute the largest source of technology transfer to Japan among all types of U.S.-Japan strategic alliances. Equity investments presumably give Japanese investors direct, first-hand access to state-of-the-art technology. The openness of the U.S. semiconductor industry to foreign investment and the appropriability of American know-how have caused the private and public sectors to be concerned about how to monitor and, where necessary, to restrict foreign investments.

A troublesome disconnect between microlevel incentives for individual U.S. firms (which want and need to attract capital) and the collective, potentially adverse, long-term impact of Japanese investments on the U.S. semiconductor industry as a whole, may result in a continuing net transfer of vital technologies from the United States.[30] What serves the interests of

---

[29]AEA Japan Office, "Japanese Electronics Acquisitions in America Since 1986" (unpublished, mimeographed listing).

[30]For an overview of the relationship between Japanese investment and technology transfer to the United States, see Committee on Japan, *Japanese Investment and Technology Transfer: An Exploration of Impacts* (Washington, D.C.: National Research Council, 1992).

individual firms may not necessarily contribute to the well being of the industry—Adam Smith's notion of the market's invisible hand notwithstanding. What is not clear and needs clarification, however, is when and under what conditions the disconnection takes place.

What makes the task difficult is the absence of an effective mechanism for monitoring, investigating, and approving foreign investments in areas deemed to be of vital importance to national security. There is no agreed-upon definition of key terms such as "strategic technologies," "national security," "economic security," or even "foreign investment." This definitional confusion and an absence of effective regulatory institutions mean that the U.S. government has no clear-cut policy to anticipate or deal with the effects on U.S. competitiveness.

Concern about the national security implications of Japanese investments prompted the U.S. government to intervene in, and halt, the proposed sale of Fairchild and other U.S. companies. It also led the U.S. Congress, in August 1988, to pass the Exon-Florio amendment to the Defense Production Act, authorizing the president to block foreign investments in strategic industries that might jeopardize U.S. national security. To establish some federal oversight of foreign investments, the president organized the Committee on Foreign Investment in the United States (CFIUS), but many observers question just how effective an agency for oversight CFIUS is. Of over 700 cases of investment reported as of June 1992, CFIUS had chosen to undertake an extensive investigation of only 14, and of the 14, it had forwarded a recommendation to block only one, the proposed acquisition of a U.S. aerospace company by mainland Chinese interests (immediately following the Tiananmen Square incident).

Although the impacts of foreign investment have great potential significance, they are multidimensional and sometimes difficult to predict. One issue is how to demarcate the concept of economic security so as to preclude stretching it beyond recognition. Another set of analytical problems relates to the absence of an accepted methodology for assessing the commercial impacts of technology transfer. These issues lead an observer to ask whether there is a real problem or whether the advantages of foreign investment balance or outweigh the drawbacks.

In looking over the AEA's list of Japanese investments in the semiconductor industry, it is difficult to tell how many are problematic in terms of serious technology loss. At first glance, one can identify a number of cases of Japanese investments that involved little or no technology transfer: ASCII-Informix, Canon-NeXT, Hitachi-National Advanced Systems, Kyocera-PictureTel, Mitsui Comtek-Raster Graphics, NKK-Silicon Graphics, and Sony-CXC. One can also cite many other cases in which important transfers did take place. The point is that neither the aggregate number of cases nor the aggregate dollar value of Japanese investments can be used as a reliable indicator of technology loss and gain. It is misleading to give the Hitachi-National Advanced Systems case the same weight as the Kubota-MIPS or Matsushita-Solbourne Computer alliances in terms of the technology transferred. It is also inaccurate to assume that all such alliances lead to a damaging outflow of critical technology.

Ideally, the data would be sorted into two simple, dichotomous categories: (1) cases in which technology is not transferred; and (2) cases in which transfer occurs. For purposes of this study, we are interested only in the second category of Japanese investment. If we could specify which types of technology qualify as "strategic" and what forms of alliances are most apt to transfer such technology, we could begin to formulate some hypotheses about the impacts of strategic alliances on the U.S. economy and technology base. In the absence of a data base that would allow us to categorize alliances along these dimensions, it is difficult for policymakers and practitioners alike to assess the extent of the problem, much less to develop effective responses.

7

Issues for U.S. Policy: National Security

The post-Cold War era is one in which the role of economic and technological factors in national security is increasing. The global changes taking place today are so sweeping that old concepts of national security must be thoroughly reexamined. In the sphere of U.S.-Japan relations, for example, one of the cornerstones of the bilateral security alliance—namely, the threat of Soviet and Communist expansion—has been undermined by the collapse of the USSR. Without the cohesion provided by a common perception of the Soviet or Communist threat, will the U.S.-Japan relationship remain intact, or will conflicts engendered by the huge trade imbalance destroy the structure of the alliance? Will the United States continue to tolerate the lopsided trade deficit? Will it permit the private sector to continue concluding alliances that give Japanese competitors access to some of America's leading-edge technology?

In the United States, the commitment to the ideology of free market economics runs so deep that government interference in private sector alliances is usually justified only in those exceptional cases in which national security is clearly jeopardized. The marketplace is regarded as the most effective mechanism for fostering economic efficiency, and government intervention—particularly protection or excessive regulatory control—is believed to distort the market and expose it to the dangerous virus of special interest lobbying and partisan politics. Given this deeply embedded ideology, the primary justification for government intervention is the nation's security, the state's highest responsibility and ultimate source of empowerment.

The problem with invoking national security is that the concept is slippery. From a strictly military point of view—territorial defense, force postures, and conflict scenarios—national security is clearly definable, but when nonmilitary factors are introduced, the concept of national security becomes exceedingly hard to specify. How are linkages made and definitional boundaries drawn? Is a certain industry—semiconductors, for example—or even a specific product such as microprocessors or DRAMs, essential to a nation's military-industrial capabilities? What is the connection between the commercial health of an industry such as semiconductors and national security? Unless answers to these and other questions can be given, the concept of national security cannot be spelled out definitively. There will always be the problem of unbounded elasticity, expanding the definitional boundaries to include almost any industry or product, be it DRAMs or aluminum, that can be argued to have an impact on national security, however indirect or marginal. It is hardly surprising, therefore, that defining national security in the post-Cold War era has become a subject of intense debate.

In years past, the issue of supply dependence and disruption was a primary focus of concern from a national security perspective. If U.S. arms manufacturers, for example, become too dependent on foreign producers of weapons components, equipment, or systems, and if the supply of those goods could be disrupted, U.S. national security would be at risk. To ensure that the worst-case scenario would never occur, the U.S. government had a relatively straightforward policy solution: developing a safety net, consisting of supply diversification (procuring from domestic and foreign sources, or from more than one foreign source) and accumulating a stockpile of supplies that could be drawn upon during emergencies.

In today's fast-paced world of high-tech weaponry and commercial competition, however, it is no longer possible to cling to old concepts of security based solely on supply disruption. The salience of economic factors has not only increased but also become intertwined with other dimensions of national security. It is more difficult today to disentangle the economic from strictly military factors and to define the concept of national security comprehensively in terms of military, political, social, and economic factors.[31] What has happened to alter the calculus of national security? Listed below are a few of the seminal developments:

1. a continual broadening and deepening of ties of international economic interdependence led by capital flows, trade, and technology transfers;

2. U.S. reliance on technological superiority (as opposed to the pure firepower of its arsenal) to maintain a credible military deterrent;

---

[31] For a thoughtful analysis, see Theodore H. Moran, "International Economics and National Security," *Foreign Affairs*, Winter 1991, pp. 74-90.

3. a shift in the directionality of innovation from the commercial spillover benefits of military-based R&D to the military spillover benefits of commercially driven R&D—in other words, a dramatic rise in the relative importance of commercial activities for national security, especially in high-technology sectors (such as semiconductors and software);

4. a steady shift in employment and output from old-line manufacturing to the high-tech and service sectors;

5. the seminal importance of being competitive in high technology industries for the nation's overall economic well being, including the productivity of manufacturing industries and the service sector;

6. the end of an era of U.S. dominance in high-technology sectors and Japan's rise as a formidable competitor (and in certain industries, the emerging leader);

7. the relative decline of American hegemony and evidence of serious and unresolved economic problems at home (e.g., budget and trade deficits);

8. the growing public perception in the United States that economic competition from Japan may pose a greater threat to America's national security than the Soviet military threat; and

9. hard questions concerning a more equitable distribution of costs and burdens associated with the provision of such collective goods as peace, stability, and security (disproportionately shouldered by the United States), and the image of Japan as an opportunistic "free rider."

Reflecting the importance of such developments, a variety of specific questions concerning the fusion of economic and military factors in the calculus of national security have arisen. Of these, the most germane are listed below:

1. How dependent has the U.S. military become on foreign vendors and on foreign technology? At what point is national security placed at risk?

2. How essential is it that U.S. corporations maintain a position of commercial leadership in high-technology industries? How would military security be affected if the United States were to lose its competitiveness?

3. Under what conditions are foreign acquisitions of domestic companies a security problem?

4. Are tags of national identity losing their meaning in this era of multinationalization and globalization? Does it matter that manufacturing facilities located in the United States bear a Japanese name, as long as the technical work force is predominantly American and the physical facilities, value added, and learning curves remain here?

5. To what extent, if at all, should the United States be concerned about the industrial policies of foreign states targeted at promoting their high-tech, strategic industries? Is it necessary for the U.S. government to

"level the playing field" by taking industrial policy measures to promote domestic producers?

6. Should foreign companies be encouraged to invest in R&D and manufacturing facilities in the United States? What are the advantages and dangers?

7. In what technological areas is leakage damaging to national security? How might it be shut off or at least contained?

The problem of defining national security interests goes far beyond simple questions of supply access and disruption; it now involves such complex issues as the commercial competitiveness of domestic producers in key high-tech industries, the scope of the state's role, and coping with the consequences of proliferating ties of economic interdependence.

How can these far-reaching and diverse issues be feasibly dealt with? In theory, the slippery notion of economic security can be best understood in terms of enhancing the twin objectives of economic efficiency and adaptability. Whatever damages economic efficiency and adaptability significantly can be construed as a threat to economic security—for example, very large budget deficits, low savings rates, inadequate investment levels in research and development, poor manpower training and education, or falling rates of productivity. In this sense, trade protection can also be considered a security threat if the costs of protection—inefficiency and structural rigidity—outweigh such benefits as the survival of an indispensable industry. Clearly the cost-benefit matrix will always involve complex calculations of difficult trade-offs. Formulating the calculus in terms of efficiency and adaptability at least provides conceptual guidelines on which to anchor what is otherwise an amorphous and unmanageable problem.

Perhaps the most parsimonious way of dealing with the relationship between military and economic factors is to funnel concerns through the prism of a fundamental question: To what extent is the nation's capacity to innovate and manufacture diminished in technologies critical to military security? In the final analysis it is the nation's capacity to generate and produce essential state-of-the-art technology that should be the overriding criteria by which definitional boundaries of military security are drawn.

Using this conceptual lens makes it possible to sort through the maze of issues that have a direct bearing on national security. If, for example, a strategic alliance involving the transfer of seminal technology, such as operational software, places America's capacity to innovate and manufacture essential weaponry at risk, then the proposed alliance would constitute a national security question requiring analysis and possible public policy action. Conversely, if the loss of commercial leadership in a particular product market, such as ASICs, is likely to have relatively limited impact on America's capacity to innovate and manufacture, then it would not be regarded as a national security problem. Even with the adoption of such a prism, a number of difficult questions remain, such as definitions of critical technologies and predictions of impacts on U.S. capacity to innovate in the long run. Focusing on the effects on innovative capability provides a foundation for addressing these complex issues.

8

Issues for U.S. Policy: Global Technological Stratification and U.S. Technological Capabilities

UPSTREAM TRENDS: THE SEMICONDUCTOR EQUIPMENT INDUSTRY

The development and growth of the semiconductor equipment industry over the past several decades give evidence of the technological stratification of the semiconductor industry. This stratification is due to the specialization of technical expertise and to the U.S. investment community's increasing preference for focus over integration. In the early days of the industry, companies were virtually self-sufficient. In the 1960s, Texas Instruments, the largest semiconductor company at the time, grew its own material, made its own packages, and designed and built its own equipment. By the late 1960s a few equipment manufacturers had started to build systems, and firms would design and specify what they wanted. In the 1970s it became clear that semiconductor firms could not remain technologically self-sufficient, as equipment makers started to suggest equipment that the firms were not intimately involved in designing such as ion implanters at National Semiconductor. The role of equipment makers became more important as the industry shifted from being labor to capital intensive. In 1967 a competitive wafer fabrication line cost about $1 million, whereas a world-class volume fabrication facility costs well over $100 million today.

By the beginning of the 1990s the semiconductor industry had come to depend on sophisticated technology in the areas of wafer manufacturing (lithography, deposition, packaging), chip design—computer-aided design and engineering (CAD/CAE), and other tools to design the electronic and process properties of the devices—and design of the systems architecture embodied in chips. Although some might argue that Hitachi possesses most of this capability, the Texas Instruments of the 1960s is gone forever, and no company can push technology forward on its own. Semiconductor equipment companies play an important role in the overall technical progress of the industry.

Technical stratification has been partially spurred by the preferences of investors. There was a flurry of investment in supplier companies in the mid-1980s, and some good companies such as Novellus were born, but interest has flagged because development cycles are typically long by venture-investing standards and because the condition of the U.S. semiconductor industry has made it a worrisome customer target. Presently, even U.S. companies that offer significant improvements on processing technology have trouble obtaining financing. One example is Hampshire Instruments, an x-ray lithography company that has faced continuing financial challenges even as technical challenges have been conquered. Thus, in general, even new U.S. equipment companies with superior technology are not current investment targets because the capital required is high, the development cycles are long, the U.S. customer base is weak, and leading Japanese customers have their own suppliers. A significant portion of the technology required for future semiconductors is not gaining new company investment support, which leaves technology development to existing U.S. companies and foreign firms.

The competitive and technical landscape differs among various segments of the equipment industry. First, in lithography equipment, where technology is tied to resist and U.S. companies invented the concepts for all leading machines, the United States has the technology for the next generation—both x-ray and "phase shift mask" with eximer lasers (today's steppers). Yet there are no commercially "leading" companies resident in the United States, the leaders being Nikon, Canon, and ASM Lithography. American entrants GCA, SVG and Ultratech are followers.

In deposition equipment, Applied Materials, the strongest U.S. equipment company, has world-class technology that must be demonstrated before it can sell its products. It has acquired a strong position in Japan through hard, long effort but has growing competition and a customer base that understands its technology to some degree in order to use it. In years past, Applied Materials would get advance payments from customers, as did other equipment suppliers, and then work with them to develop the equipment and processes necessary to build the next generation of devices. Today IBM and, to some degree, SEMATECH will support the development of a machine in this way, but the balance of the load is carried by the equipment maker.

Similar situations exist for companies in the automation and test equipment fields. They must perform equipment and process development with their own capital and then transfer their process technology to make a sale. The sales of equipment companies necessarily involve the transfer of process technology.

The processes by which component materials are developed are really no different. Makers of silicon wafers, exposure resists, masks, packages, and CAD/CAE tools must develop products with their own resources. Venture investment is not attracted to the area, again because of the general weakness of the U.S. semiconductor industry and the concern that any success in Japan will be short-lived since the technology transferred through a sale will very likely leak to local suppliers.

Although it is hard to pinpoint aggregate trends in U.S.-Japan alliances in equipment and other upstream sectors supplying the semiconductor industry, there are indications that the Japanese semiconductor equipment industry has been fairly active in U.S. investment recently.[32] A key question is whether the U.S.-owned portion of the semiconductor equipment industry can remain viable in the face of the conditions outlined above. If the answer is no, the next question is whether a predominantly Japanese-owned equipment-making industry will be adequate to support the maintenance of semiconductor process design capability in the United States.[33]

DOWNSTREAM TRENDS: SYSTEMS, COMPONENTS, AND PROPRIETARY ARCHITECTURES

On August 12, 1981, IBM Corporation launched its first personal computer, creating an industrial miniboom for semiconductors, disk drives, software, and countless other products. In executing its strategy, IBM chose to ally itself with smaller companies for critical components, notably Intel and Microsoft.

Spawned by the microprocessor, the personal computer was a triumph of miniaturization that transformed the way many people work and eroded

---

[32]See Phyllis A. Genther and Donald H. Dalton, "Japanese-Affiliated Electronics Companies: Implications for U.S. Technology Development," NTIS, March 1992, pp. 7-11. According to Japan Economic Institute and U.S. Department of Commerce data the number of Japanese-owned semiconductor manufacturing equipment plants increased from 13 in 1989 to 23 in 1990, while employment increased from 1,745 to 1,960. This "indicates growth in the supplier network of Japanese-affiliated companies . . . and suggests the replication of keiretsu structure linkages in the United States."

[33]Concerns have been raised (and denied by Japanese spokespersons) that Japanese semiconductor equipment makers may delay or deny sales of their most advanced products to U.S. companies. See U.S. General Accounting Office, *International Trade: U.S. Business Access to Certain Foreign State-of-the-Art Technology* (Washington, D.C., September 1991).

the sales of much larger machines. In fact, the opportunity in personal computers was so great that a myriad of companies sprang up around the world. In less than 10 years, the total worldwide market grew to $150 billion annually.

Times have changed, however. With the personal computer industry maturing, the growth of sales of personal computers has slowed—from 34 percent in 1988 to 10 percent in 1990. This is taking a toll on computer makers as well as on the semiconductor manufacturers who supply them.

Today, faced with price cutting, moribund profits, and an increasingly difficult future in a saturated personal computer market, the largest players in the computer industry are attempting to redefine the global market battlefield through new partnerships and strategic alliances. These alliances, along with major technological changes, will have a profound effect in determining who will be the winners and losers over the next decade in the semiconductor, software, and computer businesses. To be sure, the new personal computer alliances have vulnerabilities and face a number of challenges. Focusing the resources and coordinating the strategies of a large number of companies on a common goal is fraught with difficulties, and new technical developments or counter-strategies can raise obstacles to collaboration or rob an alliance of its rationale. Whether or not the alliances described below ultimately meet their objectives, market trends may force companies to contemplate and launch other, similar efforts.

The two alliances that have attracted the most attention are the ACE consortium and the Apple-IBM alliance. ACE was announced on April 9, 1991, and was launched by 21 industry-leading hardware and software companies. Led by Compaq Computer Corporation the alliance announced a set of specifications for an advanced computing environment. The ACE consortium's ambitious goal is the creation of a hardware and software system that can be used in everything from laptops to mainframes.

The group intends to generate compatible computers that will be equal in popularity to IBM's personal computer. The new machines will use one of two operating systems and two microprocessor chips. The two hardware architectures are based on the MIPS R3000/R4000 RISC family and the Intel 80X86 family of microprocessors.

In the months following the ACE announcement, the consortium quadrupled in size to include more than 80 computer companies from the United States, Japan, and Europe. Successful implementation of the ACE initiative would position its members to define industry direction by thwarting rivals in the commercial sector and gaining access to markets for networked computers.

If ACE succeeds, the big potential winner in the semiconductor industry would be MIPS, which licensed its RISC processor technology to several U.S. and Japanese semiconductor manufacturers. Two in particular, LSI

Logic Corporation and Nippon Electric Corporation (NEC), should benefit most from being suppliers to the ACE alliance of computer original equipment manufacturers (OEM). Furthermore, NEC's ability to supply large quantities of high-density DRAMs to the alliance members would help ensure a strong challenge in the industry.

The big potential loser in the industry would be Intel, which has enjoyed a virtual microprocessor monopoly in personal computers. If ACE succeeds, software portability across a wide range of hardware devices from a multitude of vendors would force Intel to offer more aggressive price-performance characteristics for its architecture.

During the year after ACE was formed, the consortium has run into a number of obstacles, although the members remain publicly committed to its goals. For example, Intel has accelerated the commercialization of its next generation microprocessors in response to the ACE push for a RISC personal computer standard and to the introduction of MPUs compatible with its X86 line by Advanced Micro Devices and others. Compaq decided that the performance of Intel's new products will obviate some of the rationale for RISC personal computers, and dropped out of ACE in April 1992.[34] In addition, MIPS was acquired by workstation maker Silicon Graphics in early 1992, leading some analysts to question whether all ACE computer makers would have equal access to MIPS designs. In the spring of 1992, it was unclear whether the leading companies in ACE would be able to provide strong enough leadership to push the consortium to a timely achievement of its objectives.

A second significant alliance was launched on July 5, 1991, when Apple Computer and IBM Corporation joined forces in a wide-ranging pact to jointly develop powerful, easy-to-use desktop computer systems. Formerly fierce rivals, the two companies will cooperate in the development of next-generation computer operating software and advanced RISC hardware. Furthermore, to pave the way for its joint venture with Apple, which is known as Taligent, IBM forged a series of pacts with software publishers, computer companies, and chip makers in which member companies will develop software and hardware to take advantage of the systems that Apple and IBM develop together.

Within weeks of announcement of the Apple-IBM partnership, Motorola and Siemens AG were invited to participate as key suppliers. Motorola will second source the IBM-designed RS6000 microprocessor (also known as the Power PC chip), and Siemens will produce the world's most advanced computer memory chips (at first the 16 megabit DRAM).

---

[34] Peter Lewis, "Whither the ACE Consortium? Maybe Nowhere," *The New York Times*, May 10, 1992, Section 3, p. 10.

The bold moves by these two computer giants are likely to change the ground rules for competition by forcing a redefinition of the personal computer. Codesigning technology is only the veneer on their strategy; the real purpose of the alliance is to preserve and enhance their enormous marketing clout. By deciding not to compete with each other, they have sent a strong message to Compaq, Sun Microsystems, AT&T/NCR, and numerous Japanese companies: recalibrate your strategic aspirations or face competing head-to-head against Apple-IBM.

The big potential winner in the semiconductor industry is Motorola. Long a friend of Apple and IBM, and far behind Intel in microprocessor market share, Motorola may do very well as the designated microprocessor supplier to the alliance. Another potential winner, as a group, would be Europe's struggling chip manufacturers. Bringing German electronics giant Siemens AG into the Apple-IBM camp will have two major effects. First, by sharing the burden of building a $700 million advanced chip factory in France, and sharing proven 16 megabit DRAM production technology, the partners build a base for their efforts to develop a production worthy 64 megabit DRAM. Second, because the agreement with Siemens allows for additional European partners, European computer makers will be less likely to form their own alliances in response to a perceived threat from Apple-IBM.

Whether or not Taligent can deliver what its parents intend for it is still unclear. Since the announcement of the alliance, both Apple and IBM have made moves that appear to be aimed at hedging their bets on Taligent. Still, the partners have devoted considerable resources to the venture.

The recent trends in alliances among software, semiconductor, and computer companies may well signal a shift in the balance of power in the computing industry. The key unsettled question in the 1990s is whether any alliance can protect its members from increasingly fierce competition. Some recent trends need to be examined to answer this question.

Historically in the electronics industry, as semiconductor technology becomes cheaper the prices of the personal computer or workstations using memories and microprocessors decrease. Lower prices have led to higher sales, fueling the development of entirely new applications. This remarkable relationship—cheaper memory leads to cheaper computers, which then expands the market for software—will be severely challenged in the next few years.

There is wide agreement that software represents perhaps the key battlefield for U.S. computer companies in the 1990s. The United States dominates the world's software market share at a time when software is becoming an increasing portion of value-added information technology. In 1990, the U.S. world market share was approximately 75 percent in software, 65 percent in computers, and 40 percent in semiconductors.

In the area of software, techniques such as object-oriented programming hold the promise for U.S. companies to save tremendous amounts of time and effort in creating programs from ready-made sections of computer code. On the other hand, U.S. companies worry about a new trend of several large Japanese computer companies bringing mass production techniques to writing programs.[35] Furthermore, what the Japanese companies cannot make, they can usually buy because of their capital resources. This possibility poses a severe challenge to current U.S. dominance in software.

Another area of concern for the U.S. software industry is that as applications become increasingly more powerful, they require vastly more semiconductor memory. This is especially true in computer workstations, which make sophisticated use of high-resolution graphics and digital video. The problem facing the U.S. companies is that memory production today is dominated by Japanese, and increasingly, Korean companies. Furthermore, new memory technology is, for the first time in history, becoming more costly. The real problem is that whenever memory becomes more expensive, computers become more expensive, thereby restricting the market for software.

So the problem is two-pronged for U.S.-based companies in the information industries. First, when demand is growing Japanese companies may be able to increase the price of memories and thereby slow down U.S. growth in strategic segments. Second, a slow down in growth gives Japanese companies time to catch up in software technology and thereby challenge U.S. dominance. This is one reason that both the ACE consortium and the Apple-IBM alliance included allies in semiconductor memory production—to attempt to thwart a possible restriction of memory supplies.

Another concern of U.S. companies is the trend toward miniaturization, a trend that plays to foreign strengths. In the high-growth areas of laptop and notebook computers, Japan supplies virtually all the display screens and commands a large share of the market for portable computers. Even in the new area of pen-based computers, most machines not only are manufactured abroad but require high-density memories made largely by Japan. But U.S. software strength also presents opportunities in this new environment. For example, in early 1992 a number of alliances were announced that aim to develop new products that would lie at the intersection of the computer, consumer electronics, and telecommunications equipment markets. Some analysts predict that U.S. software strength will allow computer companies such as Apple to reestablish a strong U.S. presence in consumer electronics. Still, it appears that in all of these nascent alliances—such as Apple's partnership with Sharp and AT&T's linkage with Matsushita—it is necessary

---

[35]Michael A. Cusumano, *Japan's Software Factories: A Challenge to U.S. Management* (New York: Oxford University Press, 1991).

for U.S. companies to turn to Japanese partners with consumer electronics marketing and component manufacturing capabilities.[36]

If the battlefield in the 1990s is density memories and sophisticated software, where will microprocessors fit? To be sure, the performance of microprocessors is advancing, largely as competitors pursue the RISC-design technology. Furthermore, added power is needed to run increasingly powerful applications. Yet at the same time, more and more software is being written to run independently of any particular microprocessor. This trend, if it continues, will allow any computer maker to choose any cheap microprocessor, as long as it has the horsepower to run the intended applications. Furthermore, as microprocessors enter an era of commodity pricing, a severe challenge could be posed to the profitability and future of microprocessor vendors that have relied on proprietary architectures to maintain market share. Thus, the personal computer and semiconductor industries have entered a crucial period of alignments and realignments based on global alliances. What emerges from the crucible of alliance-bloc competition will be fascinating to observe. Certainly both the processes and the outcome will shape the structure and dynamics of the personal computer and semiconductor industries (just as RISC alliances will shape the structure and dynamics of the workstation industry).

---

[36]Bob Johnstone, "Future at Hand," *Far Eastern Economic Review*, April 30, 1992, p. 74.

9

Possible Scenarios for U.S.-Japan Alliances and Their Implications for the United States

How will the trends described in Chapter 8 and other broad forces interact in the future to influence the competitive position of the U.S. semiconductor industry? How do U.S.-Japan alliances affect American semiconductor competitiveness? Do they strengthen or weaken U.S. industry? This chapter sketches several possible scenarios for the U.S. semiconductor industry, and considers factors that will determine its performance in world markets.

SCENARIO 1: GRADUAL U.S. RECOVERY

Currently, the U.S. semiconductor industry has about a 39 percent share of the worldwide market. In this optimistic scenario, the U.S. industry gradually regains a market share, rising above 40 percent share to a 45-50 percent global share. A number of conditions that would contribute to this scenario are listed in Table 8.

How plausible is this scenario? There is strong evidence that five of its conditions (items 1, 2, 3, 7, and 8 in Table 8) already exist. However, the record of U.S. semiconductor manufacturing competitiveness is mixed. Although some large corporations are making new investments and improving their processes, many smaller or older companies are reducing their plant capacity. Moreover, an increase in venture capital and long-term corporate investments appears doubtful, and although access to Japanese and Asian markets may be improving slowly, it is still marked by "buy local" practices

TABLE 8 Conditions Conducive to Realizing Scenario 1

Factor	Condition
1. Exchange rates	Declining dollar
2. Cost of capital	Lower interest rates
3. Manufacturing	Increased manufacturing investment and U.S. industry productivity
4. Investment environment	Increased venture capital and long-term corporate investments
5. Asian markets	Better access to Japanese and Asian markets
6. U.S. marketing efforts	Massive global marketing effort by U.S. companies
7. Design intensity	Design-intensive technologies increase in value
8. Standards	Success of emerging global standards (ACE, SPARC, etc.)

and structural barriers. If the conditions described improve, U.S. companies will be under less pressure to enter alliances and would have more alternatives. American-Japanese alliances will be aimed at securing long-term financing, increased access to the large Japanese market, and manufacturing capacity. Large U.S. companies would be hesitant about foregoing future opportunities, whereas small "boutique" houses would become more attractive to Japanese investors. Although alliances could help facilitate a gradual U.S. recovery, they would not serve as the main driving force. Alliances are a supplement to, not a substitute for, healthy fundamentals in the private and public sectors.

SCENARIO 2: MARKET SHARE EQUILIBRIUM

In the second scenario (status quo), U.S. worldwide semiconductor market share would hover between 35 and 40 percent, which is where it is today. Maintaining a status quo, however, would require more U.S. effort to counterbalance a greater mobilization of resources in Japan and Asia. Rising Japanese R&D and plant investments probably would not forestall growing Asian strength in DRAMs and other commodity markets, whereas U.S. companies might be able to forestall shrinkage in market share by introducing new products faster. The key factors leading to an equilibrium scenario are listed in Table 9.

In this "business-as-usual" scenario, the number of U.S.-Japan alliances would continue to form at current rates and would fluctuate with business and product cycles. There would be no pressing external reason for U.S. companies to enter alliances—beyond those already at work—and no major new pressures on Japanese companies to help U.S. companies. However, the number of U.S.-Asian alliances might increase in response to emerging

TABLE 9 Conditions Conducive to Realizing Scenario 2

Factor	Condition
1. Exchange rates	Stable dollar
2. Cost of capital	Stable interest rates
3. Manufacturing	Increased manufacturing investments and U.S. industry productivity
4. Investment environment	Stable venture capital and long-term corporate investments
5. Asian markets	Better access to Japanese and Asian markets
6. U.S. marketing efforts	Increased global marketing efforts by U.S. companies
7. Design intensity	Design-intensive technologies increase in value
8. Standards	Success of emerging global standards (ACE, SPARC, etc.)
9. Asian alliances	Use of Asian fabs expands U.S. market share

opportunities for greater cooperation in Asia, triggering defensive moves by Japanese makers to contain or counteract these rival alliances.

SCENARIO 3: GRADUAL U.S. DECLINE

Under the third scenario, the U.S. semiconductor industry would lose a small but significant portion of worldwide market share with each passing year—1 to 2 percent, for example—which may seem small for any given year but adds up to an appreciable reduction over 10 years. This would mean that America's market share would be whittled down from 39 percent to 20-25 percent by the year 2000. If this scenario were to occur, it would result from the factors listed in Table 10.

TABLE 10 Conditions Conducive to Realizing Scenario 3

Factor	Condition
1. Exchange rates	Stronger dollar
2. Cost of capital	Rising interest rates
3. Manufacturing	Inadequate manufacturing investment by U.S. companies
4. Investment environment	Reduced venture capital and corporate investments
5. Asian markets	Opportunity costs in Japanese and Asian markets
6. U.S. marketing efforts	Declining global marketing efforts by U.S. companies
7. Design intensity	Variability in value of design-intensive technologies
8. Standards	Japanese successfully adapt to emerging global standards (ACE, SPARC, etc.)
9. Asian alliances	Asian fab alliances falter

If conditions worsen as outlined, U.S. companies would feel stronger pressures to seek new Japanese ties, especially in joint R&D, manufacturing, and distribution. The erosion of America's market position would not affect U.S. companies uniformly; the impact would be most severely felt by slower-reacting companies. Opportunities would be ripe for Japanese investors to make significant investments and inroads into emerging U.S. semiconductor companies. A great deal of U.S. R&D would move offshore to Japan, Europe, and Asia in order for U.S. firms to remain competitive in regional markets.

SCENARIO 4: JAPANESE DOMINANCE

Under a somewhat worse scenario, the U.S. semiconductor industry would lose 2 to 3 percent of the global market share each year to Japan, leading to a U.S. share of only 10-15 percent in the year 2000. In looking at the underlying factors described above, this sharper decline would result from the same forces at work in the third scenario in worse form (see Table 11).

If this scenario were to come to pass, U.S. companies would have no choice but to enter into alliances with Japanese companies in order to secure manufacturing, design, market access, and global marketing and distribution channels. Alliance formation, in this sense, would emerge from weakness, not from strategic calculations. American firms would find themselves at a distinct disadvantage in negotiating with Japanese partners over the conditions of cooperation. Thus, such alliances would be likely to accelerate technology outflows. Even if Japanese firms returned state-of-the-art

TABLE 11 Conditions Conducive to Realizing Scenario 4

Factor	Condition
1. Exchange rates	Strong dollar
2. Cost of capital	Rising interest rates
3. Manufacturing	Declining manufacturing investments and lower U.S. productivity
4. Investment environment	Declining venture capital and corporate investments
5. Asian markets	Poor access to Japanese and Asian markets
6. U.S. marketing efforts	Ineffectual global marketing efforts by U.S. companies
7. Design intensity	Design-intensive technologies lost through alliances
8. Standards	Japanese successfully adapt to emerging global standards (ACE, SPARC, etc.)
9. Asian alliances	Use of Japanese and Asian fabs accelerates "leakage"

technology, U.S. industry might find itself in too weak a position to take advantage of this reverse technology flow.

SCENARIO 5: PACIFIC RIM DOMINANCE

In the last, worst-case scenario, Japanese and Asian companies would account for 70-80 percent of the world market share. American and European producers would be relegated to minority status, with only 15 and 5 percent market share, respectively. Again, in terms of the underlying factors discussed above, the conditions for their demise would include those listed in Table 12.

If this "doomsday" scenario were to materialize, U.S.-Japan alliances would have little strategic relevance, and they would not arrest the slide in U.S. competitiveness. Japanese executives would consider alliances short-term, political fixes designed to keep U.S. companies afloat financially and to minimize the inevitable fallout of anti-Japanese resentment that would ensue. The United States would be relegated to serving as merely an R&D laboratory for Asia and Japan, with little or no infrastructure for mass manufacturing. By selling raw ideas to Japanese and Asian companies, which are then able to capture the most value added from manufacturing and marketing, the United States would become in effect, a satellite participant, lacking the full range of competence in research and development, production, and marketing to be a full-time player. America would be turned into a "banana republic" in the world of high technology (see Figure 16). There

TABLE 12 Conditions Conducive to Realizing Scenario 5

Factor	Condition
1. Exchange rates	Grossly overvalued dollar
2. Cost of capital	Very high (15-20%) interest rates
3. Manufacturing	Meager manufacturing investments and sharp U.S. productivity setbacks
4. Investment environment	Marked falloff of venture capital and corporate investments
5. Asian markets	Diminished access to Japanese and Asian markets
6. U.S. marketing efforts	Woeful global marketing efforts by U.S. companies
7. Design intensity	Design-intensive technologies lost through alliances
8. Standards	Japan and Asia leverage global standards (ACE, SPARC, etc.)
9. Asian alliances	Use of Japanese and Asian fabs accelerates technology "leakage" and loss of U.S. comparative advantage in product development and standard setting

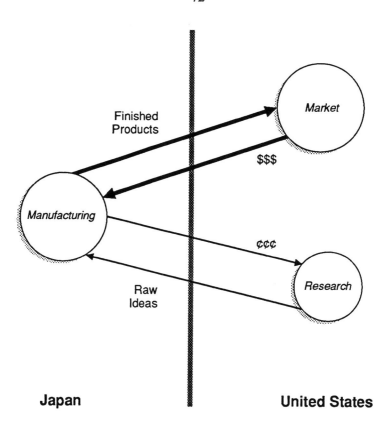

FIGURE 16 The high-tech banana republic. SOURCE: William Howard.

would be a massive loss of jobs, a traumatic "shake out" of the semiconductor industry's fragmented but dynamic structure, and an upheaval in the delicate ecosystem currently in place for the training of highly skilled, technical manpower. American engineers laid off at large semiconductor houses would have to find work at start-up "boutique" houses without fabrication facilities (financed largely by Japanese companies) or at local Japanese subsidiaries; many would find themselves out of work.

Which of the five scenarios outlined above will occur? Although accurate forecasting of the future is impossible, it appears that the U.S. semiconductor industry today stands closest to the second scenario, an equilibrium model. Whether American companies can continue to forestall structural decline following the current recession is open to doubt. Large U.S. semiconductor companies are laying off engineers and being out-invested by their Japanese competitors, which suggests the inevitability of future market share erosion unless the fundamentals of the situation can be turned around. Moreover, the entry of Japanese RISC microprocessor manufacturers and the shift to "multimedia" computing and other memory-intensive systems will enhance the market strengths of Japanese chip makers who will compete with other Asian makers for global market share. However, the implications of the shift from logic to memory intensiveness are not as yet entirely clear. New assessments in early 1992 of a U.S. semiconductor industry resurgence were tempered by questions as to whether the real trend was toward "stalemate," or "stabilization" with neither Japanese nor American producers gaining much ground.[37]

In sum, unless fundamental changes are made in the way business is done, the U.S. semiconductor industry seems likely to be headed for either Scenario 2 or perhaps the slippery slope of gradual decline outlined in Scenario 3. Of the two, Japanese companies probably would prefer to see Scenario 2 materialize; a continuation of the status quo would be less likely to aggravate potentially volatile trade tensions. To maintain the status quo, however, U.S. companies will have to pay greater attention to their competitive fundamentals, including the retention of a viable manufacturing infrastructure at home. Japanese corporations will have to do more than they have in the past to ensure a full reverse flow of technology, better access to Japanese markets, and a clearer commitment to reciprocity.

---

[37] See Andrew Pollack, "U.S. Chip Makers Stem the Tide in Trade Battles with Japanese," *New York Times*, April 9, 1992.

10

Conclusions

We believe that the alliance boom is clearly going to continue into the foreseeable future—until the computer and semiconductor industries pass their peaks of maturity and enter the stage of downward decline; even then, incentives to form alliances will remain strong. As long as the underlying forces identified in Chapters 3-5 are unchanged, the growth of American-Japanese alliances will not slow. Indeed, as foreign markets expand, commercial pressures to form alliances with companies in many different foreign countries will arise. The world semiconductor industry will see more alliances concluded between American and European firms, between Asian and American companies, and between European and Japanese corporations. In addition to the usual bilateral alliances, the number of trilateral and multilateral tie-ups is bound to increase.

As sites of the two largest semiconductor industries, linkages between the United States and Japan will probably continue to constitute the bulk of international marriages. Alliances between U.S. and Asian companies will extend beyond the United States and Japan to include new players in Singapore, Taiwan, and Korea, which now have significant capabilities in semiconductor production. Just as the types of alliance have become more varied over time, so too will the number and nationalities of corporate participants. Strategic alliances will therefore continue to function as the most important mechanism for technology transfer in the semiconductor industry well into the next century.

EXAMPLES OF U.S.-JAPAN ALLIANCES: ASSESSING COSTS AND BENEFITS

To assess the costs and benefits of alliances, it is necessary to look in some detail at specific examples. A number of examples of U.S.-Japan semiconductor alliances are treated in this report. In Appendix A, three cases are described in detail: Motorola-Toshiba, Sun Microsystems-Fujitsu, and Kubota's series of alliances with U.S. semiconductor and computer companies. Other examples are described less extensively in the section on asymmetrical pairings (Power Integrations-Matsushita, TriQuint's difficulty in forming a satisfactory alliance, and nCHIP-Sumitomo Metal and Mining) or symmetrical pairings (LSI Logic-Kawasaki Steel). Finally, in Chapter 8, examples of alliances are used to illustrate the trouble U.S. start-up companies in semiconductor equipment and materials have in attracting capital and the implications of computer industry trends for the semiconductor industry.

Together, the cases and examples are representative of the broad range of partnering firms, motivations, and mechanisms that characterize U.S.-Japan semiconductor alliances. Motorola-Toshiba is a large U.S. company-large Japanese company alliance that includes a variety of mechanisms, such as licensing, a manufacturing joint venture, joint product development, and marketing cooperation. Sun-Fujitsu is a small (at the time the alliance was formed) U.S. company-large Japanese semiconductor company linkage that centers on a supplier-manufacturer relationship comprising licensing, consigned product development, and manufacturing foundry aspects. The Kubota case illustrates the variety of mechanisms that a large Japanese company can use to partner with small U.S. firms in support of an effort to enter the semiconductor and information industries laterally. The examples of asymmetrical alliances describe the considerations of U.S. start-up companies in forming alliances with large Japanese companies.

Gaps in the representativeness of these examples are generally in areas that are important but have a limited number of examples. One fairly new type of alliance that is not included is licensing of Japanese product technology to U.S. firms. NEC and Mitsubishi Electric have licensed products to AT&T, but this pattern is still unusual. A large-scale collaborative product development effort is another type of alliance not covered in detail here, but Texas Instruments-Hitachi is perhaps the only alliance that fits this category.[38] Large government-sponsored R&D consortia, such as Japan's Very Large Scale Integration Project and SEMATECH in the United States,

---

[38] In liquid crystal displays, a type of semiconductor for all practical purposes, the DTI joint venture between IBM and Toshiba might also fit into this category. As a manufacturing joint venture, DTI is perhaps more complex than Texas Instruments-Hitachi.

are not discussed because international participation has been rare until now. However, U.S.-Japan collaboration through this mechanism is a possibility for the future, because the Japanese government has been opening consortia to foreign participation and is currently planning to launch several new semiconductor-related projects.

A closer look at alliances that have failed would also have been useful. The historical background contained in the Motorola-Toshiba case study does describe the dead ends that Motorola has encountered, but a detailed treatment of the Motorola-Hitachi or Intel-NEC alliances and the reasons they ended in court might be illuminating in light of the fact that U.S. companies such as Digital Equipment, Hewlett-Packard, and Sun are bringing a new generation of RISC microprocessors to market and licensing them to a variety of Japanese and U.S. companies. It is generally quite difficult to get access to information about U.S.-Japan alliances that have failed, particularly those that are currently under litigation.

What do these examples tell us about the costs and benefits to U.S. companies and industry of alliances with Japanese firms? Is the calculus different today than it was during the period prior to the 1980s, when American firms traded short-term income flows for technology that enabled the Japanese to build a world-class semiconductor industry? It is safe to say that the calculus is much more complicated today. A review of the experiences contained in our examples suggests conclusions about the overall costs and benefits from the U.S. perspective.

Costs

Costs of U.S.-Japan alliances can be divided into those incurred by the company that forms the alliance and those incurred by industry, including the upstream and downstream industries for which semiconductor companies are important customers and suppliers, as well as semiconductor manufacturers themselves. For individual companies, perhaps the biggest potential cost is creating a formidable competitor through technology transfer. For U.S. industry as a whole, the loss of semiconductor manufacturing capability and infrastructure resulting from a pattern of alliances may be the most serious potential cost.

1. *Transferring Enabling Technology*: Perhaps the most important technical area of the semiconductor industry in which Japanese firms are still behind the United States is the design of software codes that are embodied in advanced microprocessors. Possession of this capability would allow Japanese companies to challenge U.S. industry in its most important stronghold, and would create new competitors for a wide range of U.S. semiconductor and systems companies. The Sun-Fujitsu alliance, the Motorola-Toshiba alliance, and the Kubota-MIPS alliance involved the transfer

of microprocessor technology. In these cases it has been a number of years since the transfer took place. The Japanese partners have yet to forge ahead and become world leaders in microprocessors.

It appears that in general, Japanese firms are still working to acquire the necessary software skill base. As described in Chapter 8, microprocessing unit markets are affected by the complex interaction of compatibility, the need to have a base of applications software that can be run on an MPU, and the possibility of technical discontinuities that can quickly change the shape of particular markets (RISC). Japan's underdeveloped personal computer market is an obvious disadvantage to acquiring a critical mass of skills. The creation of competition in this area appears to be only a potential cost for U.S. companies today.

However, industry trends toward open systems may reduce the importance of proprietary MPU architectures (see discussion of downstream trends in Chapter 8). If this is the case, then the lack of capability to design superior MPUs would be less of a hurdle for Japanese companies seeking to compete in systems markets, particularly if they control the manufacture of critical components and possess sufficient marketing capability. The MPUs licensed by Sun and MIPS to Fujitsu and Kubota have allowed the Japanese partners to build manufacturing and marketing capability in advanced workstations and supercomputers. Thus far, the critical importance of MPU designs has enabled the U.S. partners to "manage" these relationships, but there is no guarantee that Japanese firms will fail to acquire the skills necessary to design MPUs or that MPUs will not decline in importance—to the benefit of Japanese companies—in the future.

2. *Transferring Incremental Technology*: The costs incurred from transferring incremental technology are less severe than those incurred by transferring enabling technology. There are no examples in this report of such costs having actually been incurred. Kubota's investment in C-Cube Microsystems and their joint development of video compression technology is, however, an example of technology transfer in an area that may have large potential in the long term, potential that Kubota is now positioned to exploit. From the U.S. standpoint, transferring the capability to develop emerging markets is an opportunity cost, yet assessing the risk in advance is inherently difficult. The actual cost depends on the technology or product achieving greater market success than predicted at the time it was licensed or acquired.

Some American critics of U.S.-Japan semiconductor alliances say that they are partly responsible for progress made by Japanese companies in markets such as gate arrays and ASICs. Perhaps the most notable example from the past that fits this general category is liquid crystal display (LCD) technology, which was developed in the United States but which Japanese companies continue to develop and commercialize. The potential market for displays in all sorts of portable information processing and consumer electronics products is enormous. The Power Integrations case is a good example of how it is increasingly possible for U.S. companies to control their technology and avoid this risk, at least in areas where the potential for a technology can be reasonably gauged. Because it has a good patent position and the courts are more inclined to protect intellectual property rights than they have been in the past, Power Integrations was able to structure an alliance in which it receives a royalty from Matsushita for internal use of the power management technology. Matsushita may be able to develop improvements on this technology that it could then utilize, but Power Integrations can keep Matsushita from selling these improvements outside (or license and incorporate them into the product) as long as the original patents remain in force.[39] Power Integrations takes some risk, but has a better chance of being able to sustain product development and stay in the race for incremental improvements of its technology.

3. *Low Return on Resources Expended*: Low return on expenditures is a short- and long-term cost that has been incurred by U.S. companies trying to break into the Japanese market. Despite the long-term organizational learning benefits described below, the Motorola-Alps joint venture is a clear example of costs incurred in attempting to break into the Japanese market. In that case, Motorola attempted to use an alliance to enter the market and avoid enabling a long-term competitor or incurring the technology transfer-related costs outlined above. Motorola finally had to give up technology to Toshiba in order to gain significant access to the Japanese market, and it has tried to structure the relationship to minimize the risk of long-term costs. The difficulty of selling in Japan has imposed long-term costs on the American semiconductor industry and may still discourage U.S. firms from investing in Japanese marketing and the technology development capability that will be necessary for long-term growth and survival.

4. *Unsuccessful Licensing Alliances*: Unsuccessful licensing alliances occur when a Japanese company uses the technology in a way that makes the American company believe the agreement is being violated. Even apart from the issue of fault in a legal sense, it is safe to say that the licensing firm in this situation regrets granting a license and typically sues. This is a short-term technology risk as opposed to the first three costs because licensing agreements that have succeeded over several years are less likely to turn acrimonious. The Motorola-Hitachi case is referred to in this report, but there is no detailed treatment of this type of case here. Two additional points should be made: (1) This risk can be incurred in licensing to Amer-

---

[39] Significant differences in the patent systems and corporate approaches to intellectual property protection in the United States and Japan (such as patent "blanketing" in Japan) must be kept in mind constantly by small U.S. companies with limited resources to expend in litigation.

ican as well as Japanese companies; U.S.-U.S. suits over microprocessors currently appear to be more common than U.S.-Japanese suits. (2) From the perspective of an American company, being in a position to license and then bring suit is preferable to not being able to enforce intellectual property protection at all, as in the case of the Texas Instruments basic patent for integrated circuits which was not granted in Japan until many years after the application.

5. *Semiconductor Specific Opportunity Costs*: There are long-term opportunity costs other than those related to particular products, which are covered in item 2. These costs can be assessed in advance, to some extent, and are incurred by individual companies and by U.S. industry as a whole. For example, when a fabless start-up decides to consign manufacturing to Japanese companies rather than build internal manufacturing capability, it passes up experience in making devices that could bring greater technical independence and could facilitate entry into other product lines. At an industry level, lost demand for semiconductor equipment and all the inputs related to process technology may result in the disappearance of U.S. manufacturing infrastructure if this is a general trend.

For a multinational corporation like Motorola, assessing the benefits and costs from corporate and industry standpoints is more complex. For the company, building process-intensive devices in Japan with Toshiba means that it gains some of the benefit of that experience for use in U.S. operations. This experience and increased access to the Japanese market put Motorola in a better position to survive as an integrated company, which will support U.S. managerial and technical jobs. If Motorola takes purposeful action to use U.S.-based equipment makers in its Japanese fabrication plants, it may largely avoid the negative implications for the U.S. manufacturing infrastructure of producing advanced devices in Japan.

6. *Foregone Synergies*: There are also opportunity costs that are more difficult to quantify because the impact goes beyond semiconductors. By definition, these costs are incurred by systems companies and industries. There are no clear-cut examples in this report, but one might speculate that if Sun had built a semiconductor manufacturing facility, it would have the incentive and the capability to develop components besides microprocessors that add value to its systems, which would constitute a competitive advantage in the future. Just as it is difficult at this point to assess the long-term benefits of Japanese capability built through linkages, it is difficult to assess the costs of U.S. manufacturing capability foregone through alliances. We can clearly see the long-term costs incurred by the U.S. economy and by upstream industries such as semiconductors because of the American exit from much consumer electronics manufacturing during the 1970s. If information processing systems take on more consumer electronics characteristics, as

discussed in Chapter 8 under downstream trends, the costs of consigning semiconductor manufacturing will likewise come into better focus.

7. *Technical Dependence*: Technical dependence is related to foregone synergies, but represents an extreme case. Reliance on a supplier or small group of suppliers for a critical component incurs the risk of supply cutoff or price gouging. Again, it is systems companies like Sun that are mainly vulnerable. Fujitsu did a great deal of custom design work on the SPARC chip, but in that specific case, Sun added the most significant value to the chip, so adequate IPR protection and competition among licensing partners safeguard Sun. In the short term, American systems makers are more vulnerable in areas such as LCDs and DRAMs where U.S. manufacturers are weak. In the long term, concentration of the most advanced manufacturing capability in Japan might give Japanese foundries more bargaining power vis à vis companies like Sun.

8. *Lost Political Independence*: Companies involved in alliances may find themselves constrained in terms of political action. A hypothetical example would be one in which a U.S. company involved in a U.S.-Japan alliance takes a position on policy matters that it would not otherwise take because of a desire to maintain good relations with the Japanese partner. Balancing the interests that relate to the alliance and those that relate to U.S. industrial and national well being may be problematic in some cases. Another aspect of this issue is not illustrated in any of the case studies, but would impact systems makers. To the extent that semiconductor alliances consign manufacturing to Japan, there is a risk that technical dependence as outlined in item 7 would serve to constrain the political positions taken by U.S. companies.

Benefits

The benefits that accrue to U.S. companies and industry as a result of linkages with Japanese companies can be largely characterized as access to resources and capabilities that allow American companies to bring products to market more quickly and effectively.

1. *Capital for Survival*: Some Kubota alliances may have ensured the survival of a small, financially weak U.S. partner. It is difficult to determine whether companies would have survived even without Japanese backing, and the Stardent bankruptcy indicates that even the deep pockets of a Japanese partner do not always prevent failure. In many cases, a U.S. start-up firm with superior technology can survive without a Japanese alliance. In the case of a U.S. start-up that would not have survived without Japanese backing, whose products do not find a significant market, benefits to U.S. industry and the economy are likely to be minimal. It may be that "capital

for survival" benefits typically accrue to the stockholders and management of the U.S. firm.

2. *Leverage of Resources for Development and Growth*: Sun-Fujitsu is a perfect illustration of the category of benefits that can accrue to American firms when an alliance is well structured. Sun was able to bring its workstation to market faster as a result of linking with Fujitsu. In this instance, the timing and access to manufacturing capability were critical, and Sun essentially leveraged Fujitsu's capabilities to build a leading position in a rapidly growing market. Such success is unusual, and requires superior technology and skills from both the American and the Japanese partners. Linkages between fabless U.S. companies and Japanese foundries yield these benefits when they are successful. In the short term, access to these resources (including investment capital and foundry services) allows the U.S. partner to build a company. This also means that U.S. entrepreneurs and engineers continue to focus on opportunities in semiconductor design, to the benefit of the engineering and chip design infrastructure in the United States. The long-term viability and contribution of design companies that do not manufacture are more difficult to assess.

3. *Leverage of Investment Resources*: Linking with Japanese partners sometimes helps larger U.S. companies leverage investment in fabrication facilities. This is largely a short-term benefit. The LSI Logic-Kawasaki Steel alliance falls into this category. For the U.S. company, the benefits of access to a world-class fabrication facility at a cost lower than it would have paid on its own are considerable. From industry and national perspectives, the benefits are less clear. The increased viability of the U.S. firm is one benefit. When the fabrication line is built in Japan, a U.S. partner may be able to contribute toward maintaining the U.S. manufacturing infrastructure by linking with U.S.-based equipment manufacturers.

4. *Leverage of Technical Resources*: American partners also benefit when Japanese partners bring technical resources to the alliance. The design work that Fujitsu performed for Sun and the DRAM process know-how Motorola receives from Toshiba illustrate this benefit, but the Hitachi-Texas Instruments alliance may be the best illustration. For the 16 megabit DRAM the firms exchanged designs, and for the 64 megabit generation they have agreed to develop a common design. To some extent the leveraging of technical and investment resources is closely related; in the Sun-Fujitsu case these benefits overlapped. These benefits are short-term in nature, but if an alliance is managed properly and the U.S. partner maintains its technical and other capabilities, continuation of the alliance and long-term benefits are possible.

5. *Access to the Japanese Market*: A short-term benefit can be turned into a long-term plus through the sustained effort of the U.S. partner to produce superior technology and to develop the capability to market in

Japan. Sun gained better access to the Japanese workstation market by linking with Fujitsu. Motorola has gained better access to the semiconductor market by linking with Toshiba. The Power Integrations-Matsushita and nCHIP-Sumitomo Metals alliances show promise for delivering market access to the U.S. partner on better terms than would have been possible 10 years ago. These alliances demonstrate that it is often necessary to trade technology for access, but a presence in the Japanese market is increasingly essential for long-term survival and growth in the semiconductor industry and in the upstream and downstream industries to which it is linked.

6. *Freedom to Focus Resources on High-Return Activities*: The Sun-Fujitsu case illustrates another benefit: rather than invest in semiconductor manufacturing, Sun was able to focus on building marketing resources and on designing the next-generation SPARC chip, which will be on sale soon. The opportunity costs and risks associated with not manufacturing, which are covered above, must be balanced against this concrete, short- and medium-term benefit.

7. *Organizational Learning*: Learning accrues to U.S. companies that use their alliances to build a capability for extracting benefits from their technical, marketing, and manufacturing activities in Japan, whether or not these activities specifically involve alliances. Motorola is a good example. The experience that Motorola has gained at considerable expense through its alliances and other long-term efforts can be utilized in the service of long-term strategic objectives. The benefits of possessing this capability can extend beyond the semiconductor industry. For example, Motorola's telecommunications equipment business is affected by government regulation, and knowledge of Japanese government operations and key officials will be useful to the company in advancing its interests in Japan's regulatory and policy spheres.

Evaluating these costs and benefits in individual cases is an extremely complex undertaking, but several generalizations can be drawn from the examples covered in this report. From the U.S. perspective, the costs of U.S.-Japan semiconductor alliances are largely—but not exclusively—long-term, potential, and difficult to quantify. Further, if the most serious costs were incurred, they would fall on actors and interests external to the company forming the alliance and would be borne in a manner that is impossible to predict. By contrast, the benefits from U.S.-Japan alliances are often more immediate, concrete, easy to quantify, and directly appropriable by the U.S. company that forms the alliance. American-Japanese alliances, like other business relationships, generally involve both risks and benefits, and require continuous adaptation to new circumstances.

Over time, the "terms of trade" in semiconductor alliances appear to be improving for U.S. participants. It is now possible to gain valuable resources, such as investment capital, access to the Japanese market, and manufacturing services and know-how. The key factors that contribute to greater bargaining power for U.S. companies are (1) competition among larger firms seeking an alliance and (2) strong intellectual property protection.

In making U.S. investments, Japanese companies have consciously steered away from the strategy of leveraged buyouts and hostile takeovers. Self-interest and the near absence of hostile takeovers in corporate Japan explain this circumspect approach; engaging in leveraged buyouts and hostile takeovers would surely ignite outrage in the United States.

Although the record of the past illustrates significant costs associated with U.S.-Japan strategic alliances, the context is changing and the examples in this report illustrate that in some cases, tangible benefits accrue to both sides. It will be important for U.S. policymakers, the American semiconductor industry, and other analysts to continue to monitor the mechanisms and impacts of U.S.-Japan strategic alliances, to assimilate experience, and to consider changes in public policy and corporate strategy where appropriate. The U.S. government and industry need to create an environment in which American firms have sufficient bargaining power in forming alliances, and the necessary information base must be developed to take full advantage of the possible benefits.

SEMICONDUCTORS AS A STRATEGIC INDUSTRY

Strategic alliances constitute an important—perhaps the most important—mechanism for semiconductur technology transfer between the United States and Japan. Although these alliances are normally seen as instruments of technology transmission, they also reflect deep-seated forces at work within the two economies. We believe that it is unrealistic to expect alliances to correct structural deficiencies in America's own manufacturing infrastructure (even though U.S. companies ought to insist on a reverse flow of manufacturing know-how from Japan to strengthen the U.S. infrastructure); nor is it realistic to expect alliances to change the short-term time horizons of U.S. management. Strategic alliances are manifestations of underlying conditions, not the root cause of what ails the U.S. semiconductor industry.

In part, the erosion of U.S. manufacturing capabilities is a reflection of changing comparative advantage and the attractiveness of shifting into the higher value-added segments of the semiconductor industry—specifically into software, microprocessors, and customized or semicustomized areas. Whether it makes long-term sense for the U.S. semiconductor industry as a whole to abandon mass manufacturing in stepping up the ladder of value added is cause for concern. Can commodity markets be abandoned without losing control of process technology, important revenue streams, key components, and strategic flexibility for downstream applications? If the history of the U.S. automobile and machine tool industries offers any lesson, it is that the costs and dangers of neglecting manufacturing must be taken seriously.

Although focusing on improving the outcomes of strategic alliances will not, by itself, provide a panacea for the problems of U.S. industry, *this study suggests the need for a conscious effort to structure the alliances to ensure reciprocal and long-term benefits to the United States.* This is primarily a challenge for the private sector, particularly U.S. companies, but in a number of ways government policy could have a positive impact. As a prerequisite, it is necessary to have a clear sense of what alliances bring benefits to the United States and, therefore, should be encouraged. Alliances are growing more complex, but this study has suggested some essential elements of advantageous alliances (see Table 13).

Generally speaking, these alliances require ongoing interaction, considerable investment of resources, more costs and risks, and management challenges, but the rewards are also greater.

The worst alliances, from a U.S. perspective, have been one-shot transactions that mortgage the company's and America's future by selling its more valuable, state-of-the-art technologies without bringing a reverse flow

TABLE 13 Elements of Advantageous Alliances

Require, as a prerequisite, thoroughgoing analysis and careful preparation—knowledge of the strengths and weaknesses of the partner and of the value of technology and other assets, as well as a clear set of objectives;

provide mechanisms for accessing Japanese technology, particularly manufacturing technology, and for strengthening U.S.-based production;

open opportunities for U.S. firms to sell their products in Japan;

provide intellectual property protection for key technologies;

balance the complementary assets and resources of the Japanese and U.S. partners;

provide mechanisms for flexible adaptation and periodic review of the immediate impacts and possible longer-term implications; and

bring together company interests and those of the nation (by enhancing value-added production and technology diffusion through training in advanced engineering, and constructive relationships with U.S. universities, R&D consortia, and other firms in the industry).

of benefits in return (through technology transfer to the United States or the enhancement of a U.S. company's market position in Japan).

The working group believes that the semiconductor industry, a vital upstream segment of the crucial information industries, is a "strategic industry" essential to the nation's well being. The reasons and criteria for labeling semiconductors a strategic industry, though seldom spelled out, are (1) the importance of semiconductor components for superior performance in military hardware; (2) the centrality of semiconductor technology for achieving breakthroughs in computers and information-based high technology; (3) the broad range and versatility of semiconductor applications for large end user industries, including automobiles and the service sector; and (4) the pivotal position of the semiconductor industry in the system of institutions and practices underlying America's capacity to innovate—employment, manufacturing infrastructure, R&D, graduate training, and so forth. The case for semiconductors rests on nothing less than their importance in maintaining America's *capacity to innovate and commercialize technology*—and all that this implies for military, industrial, and technological leadership.

Strategic importance is time bound. No designation is immutable, and in the case of semiconductors, the leeway to undergo continuous technological change, permitting major new product breakthroughs in computers, telecommunications, and consumer electronics, is not going to last forever. If and when the technological and commercial growth trajectories level off or fall, the rationale for calling semiconductors "strategic" will disappear. For the moment, however, and into the foreseeable future, the case is compelling.

Because the semiconductor industry is strategic, a number of imperatives should be considered by the industry, government, and universities.

1. It is essential that the United States maintain a strong, efficient, and world-competitive semiconductor industry.

2. To remain a world-class competitor, the U.S. semiconductor industry needs to maintain the full complement of capabilities, including leading-edge R&D, fabrication, equipment making, manufacturing, some testing and assembly, marketing, and servicing.

3. American producers must not abandon key product markets with respect to technology (such as mask alignment) or revenue flow (e.g., DRAMs and microprocessors).

4. Since volume production and sales across a number of product markets will separate the front-runners from the rest of the pack, the U.S. semiconductor industry cannot afford to rely exclusively on the computer industry to drive its growth; it must also ride the wave of growth in consumer

electronics (including high definition television), telecommunications, aerospace, and other end-user industries.

5. Successful penetration of the Japanese and other rapidly growing Asian markets is imperative.

6. If structured properly, strategic alliances with Japanese companies can contribute to the maintenance of manufacturing and fabrication facilities in the United States and to the expansion of market opportunities abroad.

If the vigor of the U.S. semiconductor industry is important to the national interest, what specific responsibilities fall on the U.S. government and the private sector to preserve and enhance this national interest?

COMPETITIVE ADVANTAGE: ISSUES FOR U.S. INDUSTRY

1. One of the main benchmarks—and long-term objectives—of the U.S. semiconductor industry should be not simply the crossing of a market share threshold in the Japanese market (for example 20 percent in any given year) but, more fundamentally, the establishment of a permanent foothold in Japan's industrial structure, a breaking into the labyrinth of long-term, interfirm relationships. In particular, U.S. producers must position themselves to get involved at the design-in phases of end user applications.

2. Strategic alliances, effectively structured, constitute perhaps the best means of establishing a permanent foothold in the Japanese semiconductor market.

3. Strategic alliances should be used to adapt to the rapidly changing commercial environment and to advance adaptive efficiency in the utilization of resources.

4. Concretely, the prime objective should be to upgrade America's manufacturing and equipment-making capabilities. The continuing erosion of this infrastructure could undermine America's capacity to maintain a world-class semiconductor industry, and special efforts may be required by industry in partnership with government to meet this goal.[40]

5. American firms should make it a high priority to obtain a reverse flow of technology in forming strategic alliances, particularly in the area of manufacturing know-how. The construction of state-of-the-art fabrication facilities in the United States, which employ and train U.S. workers, could help to meet these goals.

6. By strengthening America's own manufacturing infrastructure, U.S. firms can also deal from positions of greater strength in strategic alliances; this will have the effect of making the flow of technology better balanced.

---

[40]One promising trend that might be encouraged is the emergence of contract manufacturing service companies such as Solectron, which recently won the Malcolm Baldridge quality award.

7. The semiconductor and related industries should analyze and communicate lessons learned from past experiences in strategic alliances with Japanese firms; the diffusion of knowledge should help companies with no prior experience; industry associations could establish a forum to accomplish these goals of diffusing organizational know-how; firms should work with government to foster these goals.

8. Any U.S. firms contemplating alliances with Japanese companies should enter with the idea of learning as much as possible (from government and other U.S. firms) and of effectively applying what they learn to the development of new sources of competitive strength. In the past, a glaring asymmetry existed: too many U.S. firms entered with only short-term, quick-fix objectives in mind, whereas their Japanese counterparts approached the alliance in a spirit of moving down an organizational learning curve and creating long-term competitive strengths.

9. To get the most mileage from strategic alliances with the Japanese, many U.S. firms will have to reexamine and revamp their organizational structures and strategies; this may require the establishment of new incentives to nurture individuals who can work effectively with Japanese counterparts and improve capabilities to develop technology-based strategies.

NATIONAL INTERESTS: ISSUES FOR THE U.S. GOVERNMENT

1. Sound macroeconomic policies are a necessary but not sufficient condition for world-class competitiveness in semiconductors. Serious consideration should be given to measures to rectify America's low savings and investment ratios, deficits, including revamping the tax framework in the United States.

2. Because U.S. institutions and procedures are not well suited to formulate and implement comprehensive industrial policy, the U.S. government must be careful not to do long-term damage to the semiconductor industry or related end user industries by distorting the incentives to be efficient.

3. In specific areas judged to be critical for national security or other reasons (such as semiconductor equipment manufacturing), government policy can have a positive effect in maintaining technological capabilities within the United States and can provide incentives for "advantageous liaisons" between U.S. and Japanese semiconductor companies. Department of Defense support for SEMATECH should be continued. Policymakers could also consider measures that encourage the U.S. semiconductor industry to invest in capital equipment at a level that will maintain a vibrant manufacturing and process design infrastructure in the United States.

4. Protection of U.S. semiconductor markets should be avoided in prin-

ciple, except when foreign practices clearly violate international rules and domestic laws (e.g., dumping and predatory pricing) or when national security is clearly jeopardized. Brandishing the threat to close the U.S. market as an instrument of diplomatic leverage or retaliation against bilateral trade imbalances ought to be resisted.

5. The United States should develop a variety of approaches to strengthening semiconductors as a strategic industry through incentives for R&D, for companies that export, and for alliances that feature the transfer of advanced manufacturing technology to the United States. Such incentives might include preferential access for foreign-based firms involved in "advantageous alliances" to participate in government-supported R&D in the United States.

6. The U.S. government's pressure on Japan to open its semiconductor market has been instrumental in helping American companies break through the structural barriers of Japanese industrial networks. As long as U.S. firms encounter impediments, appropriate pressures should be continued.

7. In principle, flows of Japanese investment into the United States should not be obstructed unless such investments can be shown to pose serious dangers to competition, to maintenance of a U.S. production base in key industry segments such as semiconductor equipment manufacturing, or to U.S. security interests. Foreign capital has contributed to the development of many small U.S. firms and start-up companies.

8. The U.S. government (in cooperation with the private sector) should improve mechanisms to collect and analyze information on investment and technology flows, and efforts should be made to assess their immediate and likely impacts. This information and analysis should be made available in a timely fashion to the public and to policymakers involved in the Committee on Foreign Investments in the United States (CFIUS). The U.S. government could work with industry to establish a West Coast facility to provide advice and information to U.S. firms contemplating strategic alliances.

9. Federal funding in support of basic research and graduate training in semiconductor-related fields at universities should be increased, and efforts should be made to encourage cooperation between universities and industry.

The specter of the worst-case scenario (the United States supplying new ideas to foreign manufacturers who derive most of the value added) is a real concern. This study of U.S.-Japan technology linkages in semiconductors suggests that considerable effort will be required to avert this outcome. Ensuring the long-term competitiveness of the U.S. semiconductor industry and of the United States as a place of production requires not only economic revitalization at the macroeconomic level, but also strategic alliances structured to produce positive, demonstrable benefits to the United States. If U.S.-Japan alliances can bring in vital technology, manufacturing know-how, and resources—instead of serving primarily as conduits for technology transfer to Japan—the long-term interests of both countries will be served.

The role of strategic alliances has grown enormously over the past decade and will continue to increase. Whether or not such alliances expand the overall pie and lead to an equitable distribution of benefits depends largely on the ability of the participants to make the most out of them. Their success will have important implications not only for the participants but also for the technology base, the economic well being, and the national security of the United States as a country.

Appendix A

Case Studies of U.S.-Japan Technology Linkages in Semiconductors

Several case studies of U.S.-Japan strategic alliances are presented here. A range of alliance types are examined, including a successful joint venture, an American start-up company that leveraged its semiconductor alliance with a Japanese partner to establish a strong market position in workstations, and a Japanese manufacturer that is employing alliances with U.S. start-ups as a key part of an overall diversification strategy.

CASE I: MOTOROLA-TOSHIBA

Since late 1986, the semiconductor alliance between Motorola and Toshiba has evolved in the direction of deeper interaction and a broader range of collaborative activity. The alliance is sometimes cited as an example of a U.S.-Japan linkage in which the technological benefits to the partners have been relatively evenly balanced. Although commitments to confidentiality by both sides necessitate a somewhat speculative treatment of the technology transfer mechanisms and impacts, the evolution of the relationship to date indicates that it has worked for both companies. The strategic orientation of the firms appears to hold out the promise of further mutually beneficial interaction in semiconductors and in other areas.

Historical Background:
Motorola's Approach to the Japanese Semiconductor Market

From Motorola's viewpoint, the conclusion of its alliance with Toshiba should be seen in the context of the company's effort over several decades to gain access to the Japanese semiconductor market. From the late 1960s through the mid-1980s, Motorola shifted tactics a number of times in response to changes in Japanese government policy, market conditions, and lessons that it learned as a business organization.

In 1962, Motorola set up a sales office in Japan, but it soon became clear that this approach would not lead to significant participation in the Japanese market. At that time, Japan's formal trade barriers were substantial: the Ministry of International Trade and Industry (MITI) and the major electronics manufacturers tried to leverage protection of the domestic market and government-sponsored R&D programs in an effort to break the worldwide dominance of U.S. manufacturers in the semiconductor and computer industries. Although the success of this strategy became apparent a decade later, in 1970 the outcome was still in considerable doubt.

Motorola would have liked to set up a manufacturing facility in Japan to get around these trade barriers and participate in the rapidly growing Japanese market, which was led by the booming demand for televisions and hand-held calculators. However, Japanese government policy also restricted foreign direct investment. In order to manufacture, a joint venture with a Japanese partner had to be formed, with no more than 50 percent of the venture owned by the foreign entity. Motorola set up a joint venture with Alps Electric Co., Ltd., an electronic component maker with which it had previous dealings in other product lines. The joint venture, Alps-Motorola Semiconductor K.K. (AMSK), did back-end assembly and testing of devices, Japanese sales, and warehousing.

Motorola hoped that AMSK would evolve into a major Japanese supplier of its products. The venture ran into a number of difficulties and collapsed during the 1974-1975 industry recession. Besides the business climate, this failure may have been partly due to Motorola's inexperience in operating in Japan. The original joint venture agreement was not entirely explicit about spheres of authority, an omission that might not have made a difference in another setting but led to problems in Japan. Although the venture failed, those connected with it learned to appreciate the complexity of doing business in Japan, the value of making agreements with partners as explicit as possible, and the effort and care that would be necessary to have a chance at success. Motorola also learned that it was probably best to leave personnel and other administrative functions in the hands of Japanese partners while retaining authority over operational matters.

For several years, Motorola went back to operating a sales office as its primary presence in the Japanese semiconductor market. In the late 1970s, an engineering office was opened to establish closer relations with customers, and this office took on some testing and quality control functions. The sales force and most of the management were Japanese, yet market share was stuck at a low level. It came as no consolation that Motorola's American rivals, such as Texas Instruments, were also not doing very well in Japan. Around 1980, Motorola realized that the continuing growth of the Japanese industry necessitated another focused effort to gain access and to develop a manufacturing capability. A semiconductor company without a significant foothold in Japan would be missing a large chunk of the world market. In addition, Japanese customers were the most demanding in the world, and lessons learned from marketing in Japan could be applied to the benefit of global operations.

Motorola decided to try another joint venture, choosing Toko, Inc., as its partner. Toko was a small electronic component maker—primarily of parts for radios—that was not tightly bound to any of the large *keiretsu* groups. Most Japanese restrictions on wholly owned foreign subsidiaries were being lifted around 1980, so the decision to try another joint venture was based on the desire to overcome an outsider image that might hinder hiring and other local activities. Toko was "outside the loop" of relationships with MITI and large banks enjoyed by the major electronics companies, and was actively seeking an American partner. The interests of the two companies were complementary, and the alliance lasted for about two years, after which Motorola acquired the joint venture and renamed it Nippon Motorola.

The subsidiary does wafer fabrication and packaging for the Japanese market, and is also the world supplier of several Motorola products. Nippon Motorola made a major effort to meet the needs of the Japanese market, but progress was slow and difficult. Some marketing staff perceived that in a number of cases in the early 1980s, clients bought devices from Japanese companies after Nippon Motorola had done the application work.

Motorola also concluded several licensing agreements covering the 68000 series of microprocessors with Hitachi, Ltd. in the early 1980s, but this alliance ended in acrimony and lawsuits as Motorola and Hitachi accused each other of patent infringement. After a court decision that found fault on both sides, the companies reached a cross-licensing agreement.

Motivations and Negotiating Issues

In the mid-1980s, Motorola realized that it needed to change tactics again to strengthen its position in Japan. A partnership with a major Japanese electronics maker or trading company would give Motorola access to more of the Japanese market. Motorola's Japanese competitors had all become quite large, and large American companies with significant capability in microelectronics, such as RCA, seemed to be disappearing.

The decision to forge an alliance with a Japanese company was not taken lightly; some Motorola managers had hard feelings about past experiences in Japan. Yet by influencing Motorola toward a cautious, deliberate approach, this experience may have played a role in the eventual success of the Toshiba partnership. Before approaching any companies, Motorola evaluated possible Japanese partners by looking at both business and nonbusiness factors: Toshiba rose to the top. Its partnerships with foreign companies had generally gone well, with both sides benefiting. It was judged that Toshiba could protect proprietary technology from other Japanese companies, and that Motorola could collaborate with Toshiba and not create a more serious competitor down the road. Some have asserted that the exchange of Motorola's microprocessor technology for Toshiba's DRAM design and manufacturing technology is the heart of the relationship. However, it is important to point out that Motorola, in initiating the negotiations, chose the partner that had a compatible culture and could provide the best market access.

For Toshiba, the partnership was attractive because of Motorola's leadership in microprocessors. Toshiba moved past NEC to become the leading Japanese maker of DRAMs in the mid-1980s, but it lagged behind NEC and Fujitsu in microprocessors. An alliance with Motorola giving Toshiba access to microprocessor technology would help redress that deficiency.

Talks between the two companies went on for more than a year and involved technical, business, and marketing personnel. The first element of the partnership announced in the press was an agreement for Motorola to purchase DRAM dice fabricated by Toshiba in August 1986. In December of the same year, the two companies announced an agreement to engage in negotiations to extend the relationship further.

The negotiations were conducted against the backdrop of a severe industry slump, particularly in DRAMs, and U.S. government antidumping actions against the Japanese semiconductor industry. The antidumping actions led to the conclusion of the U.S.-Japan Semiconductor Agreement in September 1986. Motorola had stopped making DRAMs in the United States for merchant sale at the end of 1985, as had all U.S. merchant companies except Texas Instruments and Micron Technologies. Although Motorola played a role in setting U.S. industry policy toward the U.S.-Japan semiconductor talks, the political atmosphere did not have a direct impact on negotiations with Toshiba.

Initial Structure of the Linkage and Technology Transfer

The basic elements of the partnership were set in laying the groundwork for the manufacturing joint venture, and reflected both Motorola's need for access to the Japanese market and Toshiba's desire for microprocessor technology. The agreement specified an equation whereby Toshiba would be given phased access to Motorola microprocessor know-how as Motorola reached given levels of market share in Japan. The market share measurement included all semiconductor product lines, not only the metal-oxide semiconductor (MOS) products manufactured by the joint venture facility.

There were a number of advantages to this form of agreement for Motorola. First, jointly building and managing a wafer fabrication facility committed both sides to a long-term investment of financial and human resources. Second, the arrangement contained built-in incentives for success beyond the desire to see a return on investment. To the extent that Motorola's Japanese sales rose, Toshiba would gain access to microprocessor know-how and the rights to utilize it. Both sides would suffer from a poor facility or if the personnel assigned to the partnership were not first rate. Third, by managing technology transfer within the framework of the joint venture, Motorola would have more control over the transfer process. Fourth, as an addition to the basic trade of technology for market access, Motorola would gain access to DRAMs manufactured at the facility, which it was no longer making in the United States, and could offer its customers a full product line. Finally, Motorola could access Toshiba's DRAM manufacturing know-how for use in its other fabrication facilities. The partners designated a site for the joint venture, Tohoku Semiconductor, in May 1987. The facility, which is located in Sendai, Miyagi Prefecture, was completed about a year later. Motorola began by transferring know-how concerning its 8-bit microprocessor, with an understanding that the technology transfer would be extended to its more advanced products as the sales milestones were hit. Although there were indications that the partnership hit some rough spots in 1988 and 1989, Motorola's sales in Japan increased markedly, especially at the beginning.

The microprocessor technology transferred to Toshiba through the alliance consists of circuit design and software operating codes. Toshiba became an original equipment manufacturer (OEM) for Motorola's 68020 microprocessor (a 32-bit product) in 1988. This agreement signified an upgrading of the relationship from licensing basic technology in phases and probably followed the achievement of one of Motorola's sales goals. In May 1989 the two companies agreed to expand cooperation at the Sendai facility to 4 megabit DRAMs. In June 1990 the two companies announced plans to codevelop a microprocessor for a new Toyota engine. That Motorola and Toshiba would be comfortable enough to work together on design indicates that a significant level of mutual trust had been established by that point.

The technological significance of the elements of the alliance related to memories is more difficult to evaluate. The strategies and capabilities of Japanese companies have, over the past decade, made the manufacture of DRAMs a "game" that depends on high capital investment expenditure in order to incorporate the latest developments in process technology and reach high volumes of production quickly. The firms that are ahead of the curve and can scale their production up quickly will attain lower unit costs than competitors. Motorola has been able to obtain access to a supply of DRAMs from Tohoku Semiconductor to fill out its product offering, and has been able to stay abreast of process technology improvements, which are driven by DRAMs, and to incorporate these in its other facilities.

Business Impacts and Future Prospects

The transfer of complementary technologies does not ensure benefits on the bottom line. The Motorola-Toshiba case is instructive because there are clearly areas in which both sides have benefited, but in other areas it appears that significant benefits have yet to materialize. The relationship is so complex that it is necessary to divide the consideration of impacts and future prospects into several parts.

1. Marketing Collaboration

Motorola appears to have made important progress in cracking the Japanese market since teaming with Toshiba, which was its basic goal (see Table 14). At the outset of the partnership, Toshiba had marketing objectives as well, but this element was not a major consideration and has not assumed importance since.

From the viewpoint of the U.S. industry, access to the Japanese semiconductor market has been an inherently political problem. Even after formal trade barriers were lifted, the established relationships between sup-

TABLE 14 Motorola's Semiconductor Sales in Japan (estimated)

Year	Amount ($ million)	Change from Previous Year (%)
1984	106	NA
1985	73	−31
1986	113	55
1987	158	40
1988	273	73
1989	361	32
1990	436	21
1991	497	14

NOTE: Includes all products carrying the Motorola brand and sold in Japan, regardless of country of origin.

SOURCE: Motorola, Inc.

pliers and manufacturers—some say involving informal understandings between the major firms to divide the labor in low-demand devices—have prevented U.S. firms with competitive products and prices from selling to auto makers, consumer electronics makers, and other major users of chips. The Toshiba partnership has worked for Motorola in overcoming some of these obstacles. The joint development of a microprocessor for Toyota is a good example. Both Toshiba and Toyota are members of the corporate group centered on Mitsui Bank (now Sakura Bank), and there is a large potential market for Motorola products among companies with which Toshiba has close relations.

Helping Motorola crack the Japanese market also helps Toshiba increase the U.S. semiconductor content of its own products and those of group companies; it can show the U.S. government that it is doing its best to reach the 20 percent foreign chip share targeted by the 1991 U.S.-Japan semiconductor agreement. Reports in the Japanese press that appeared in autumn 1991 indicate that the firms are planning to move to a higher level of marketing cooperation by mutually second sourcing their products, which should open up even more business for Motorola.

In 1990, Toshiba and Motorola agreed to collaborate on developing semiconductors relevant to high-definition television (HDTV). The history and implications of this particular part of the alliance illustrate the complex interplay among technological, business, and political motivations and benefits that may come to characterize a number of U.S.-Japan linkages in semiconductors and perhaps other industries.

The first report of Motorola's attempt to establish "design-in" relationships for HDTV chips came in May 1989 in the Nihon Keizai Shimbun. Motorola President George Fisher raised the issue on a trip to Tokyo with the director of the Machinery and Information Industries Bureau at MITI. The agreement with Toshiba to expand their relationship in this field came about a year later. Concrete results have yet to materialize, so discussion of the implications is necessarily somewhat speculative.

Although high definition television (HDTV) sets are still very expensive and the amount of broadcast time remains limited in Japan, the potential Japanese and world markets are huge even if Japan, the United States, and Europe adopt different standards. The benefits of establishing ground-floor supplier relationships with HDTV set manufacturers may prove to be significant for chip makers. Toshiba might also reap substantial benefits. Toshiba is generally considered to be behind some of its Japanese competitors, such as Matsushita and Sony, in advanced consumer electronics. Motorola may give Toshiba a boost as a key supplier.

The most significant potential benefits would come down the road in connection with the opening of the U.S. market to advanced television. Motorola may play an important role in developing microelectronics that implement whatever HDTV standard the U.S. Federal Communications Commission (FCC) finally decides on (in mid-1992). If Toshiba partners with Motorola to develop chips for sets directed at the U.S. market, it may gain a jump on its Japanese competitors, and will also gain politically by having higher U.S. components content. Both companies have complementary capabilities to influence the development of HDTV in the Japanese and American markets. To the extent that the market for HDTV or other memory-intensive intermediate products such as "improved" or "enhanced" definition systems grows in coming years, the demand for DRAMs will also grow, improving the prospects for that facet of the relationship (see item 3 below).

The ramifications of cooperation between Motorola and Toshiba in semiconductors for HDTV go beyond the two companies. In late 1991, some months after the Motorola-Toshiba collaboration in this area was announced, several similar consortia to develop HDTV chips were announced by Japanese companies. Each included at least one U.S. semiconductor company that had long-established linkages with Japanese partners. They were interpreted as being partly motivated by the Japanese government's desire to increase the share of foreign semiconductors sold in Japan in the face of the rapidly rising trade surplus. A Japanese press report on one of these consortia stated that the U.S. partner was not expected to contribute technically to the project but was recruited—perhaps at the suggestion of MITI—solely out of the need for an American name. Still, the final result is an interesting pattern in which Motorola identified a business opportunity, touched bases with MITI, and then negotiated an agreement with its Japanese partner. Subsequently, MITI and the major Japanese firms used the Motorola-Toshiba agreement as a model for similar design-in arrangements, partly for diplomatic reasons.

2. Microprocessors

Motorola has transferred technology to Toshiba connected with its 68000 series of microprocessors. Toshiba may be using this exposure to build more expertise in the microprocessor area but thus far has not made a great deal of progress in the market. As in the United States, the main competition in Japan is between Intel's X86 line (in the personal computer market) and the RISC microprocessors of Sun Microsystems and others (in engineering workstations and, increasingly, personal computers). Even in its own end products, Toshiba uses a variety of microprocessors. The Sun SPARC chip is used in Toshiba's laptops, and Toshiba has recently signed licensing agreements with Sun and IDT for their new designs.

This lack of success is related to the trouble Motorola has had in breaking Intel's dominance in the personal computer market. By most accounts a technically superb chip, the 68000 is used in the Apple Macintosh but in few other high-demand personal computers. Motorola's early efforts to license the 68000 to Hitachi and Philips/Signetics did not give it wide enough breadth, and the Hitachi license ended in a lawsuit, as mentioned earlier. Motorola microprocessors have enjoyed more acceptance in engineering workstations running the Unix operating system. The movement toward open systems, RISC, and other technical trends may give the microprocessor market more of a commodity character in coming years, in which case Motorola's failure to achieve a clear victory over Intel might not prove fatal in the long-term. However, it may leave Toshiba without significant short-term benefits to show from this aspect of the alliance.

3. DRAMs

The DRAM business requires the ability to invest—at great risk—large amounts in capital equipment as well as in developing superior technology. Motorola may derive significant benefits from exposure to Toshiba's production technology because it can put lessons learned at Tohoku Semiconductor to work in its U.S. and European fabrication facilities. This may allow it to leverage the huge investments necessary to stay in the DRAM manufacturing technology race. Motorola continues to make DRAMs in Europe, and press reports over the past year and a half have indicated that Motorola planned to reenter the DRAM business with a new fabrication facility in Texas, and that Toshiba and Motorola were planning to join forces to build a new European fab.

However, as of 1991, these plans appeared to have been put on hold—as were other DRAM ventures previously announced by various companies and consortia—because a long industry slump showed no sign of ending. Although there have been reports that leading DRAM makers, such as IBM and Toshiba, are focusing their efforts on launching the 16M DRAM earlier than the traditional DRAM product cycle would indicate, industry experts say that this is not occurring. Plans for a Motorola-Toshiba European fabrication facility have been canceled. It now appears that Toshiba, rather than build its own European fabrication facility or build one jointly with Motorola, will assemble and test devices fabricated at Motorola's European facility for the time being. Interestingly, there has been no announcement at this point that the Motorola-Toshiba venture will extend to the 16 megabit DRAM.

As in the case of microprocessors, circumstances beyond the control of the partners have led to less concrete benefit from the DRAM technology transfer than might have been expected at the outset. However, the firms can implement their individual and joint investment strategies when market conditions improve, as they undoubtedly will. In this way, they can spread the considerable expense and risk presently attached to investment in DRAMs. In the meantime, Motorola obtains technological benefits, although these are difficult for outsiders to quantify. As in the marketing cooperation area, Toshiba will benefit politically and perhaps on its bottom line by teaming with Motorola in its bid to "Europeanize."

Implications

The Motorola-Toshiba alliance has had mixed results in some business areas, but the relationship exhibits a logic that extends beyond short-term business results. Ironically, it is not the heralded swap of microprocessors for DRAM technology that has brought the most significant benefits, but the marketing collaboration.

What of the future? Both firms are strategically sophisticated and long term oriented. There are two areas in which this linkage may develop in interesting ways in coming years: complementary approaches to systems and the treatment of political issues.

One market that will probably grow in coming years is that resulting from the fusion of cellular telecommunications and portable computing. Motorola's technical strengths in the former—including efforts to develop a global cellular network linked by satellites—and Toshiba's in the latter point to promising possibilities for collaboration, but to date no agreements have been announced in this area.

As for future political implications, Michael Borrus sets forth one view of U.S.-Japan alliances, such as that of Motorola-Toshiba, by remarking that "there is a widely held belief that, through the relationships, U.S. firms can be made increasingly dependent on their Japanese partners for technology and markets to the point that U.S. firms sacrifice their autonomy of action."[41]

There may well be a long-term danger of this type for Motorola or other U.S. companies with close linkages to Japanese firms. The surprisingly accommodating stance displayed by the U.S. semiconductor industry during negotiation of the renewal of the U.S.-Japan Semiconductor Agreement in 1991 was seen as evidence of this dependence by some.[42] For Motorola itself, however, autonomy of action does not appear to have been impaired. For example, Fisher's call on MITI concerning HDTV chips in 1989 came several months after the resolution of a contentious dispute in which the United States Trade Representative tangled with the Ministry of Posts and

---

[41] Michael Borrus, "Chips of State," *Issues in Science and Technology*, Fall 1990, p. 46.

[42] This may have been more the result of pressure from U.S. users than from Japanese partners; at the time of the negotiations, dumping had ceased and progress was discerned in overall U.S. design-ins and sales as indicative of long-term market share growth.

Telecommunications over cellular phone standards. The dispute was resolved in Motorola's favor. In 1990, Motorola received Japan Development Bank financing for the construction of a completely Motorola-owned assembly and test facility in Sendai, the first 100 percent foreign capital firm to receive such loans. Also, in the autumn of 1991 the Japanese press reported that former Motorola chairman, and now chairman of SEMATECH Robert Galvin was the guiding force behind an effort to build cooperation between SEMATECH and the European microelectronics consortium JESSI—a move interpreted as being targeted at Japan.

It appears that, thus far, Motorola is leveraging the marketing and political influence of its Japanese partners, Toshiba in particular, to increase its presence in Japan while retaining the independence to take positions critical of Japan when this serves corporate interests. A great deal of Motorola's leverage in its relationship with Toshiba undoubtedly stems from U.S. pressure on Japan to open its economy, in particular the semiconductor industry. It would not be in Motorola's best interests to lose this leverage by doing the bidding of the Japanese government in Washington.

A critical lesson of this linkage is the importance of persistence for U.S. companies. To reap the benefits from its relationship with Toshiba, Motorola has had to spend many years of building its presence in Japan, doing its homework, and knocking on doors that remained closed until fairly recently.

CASE II: SUN-FUJITSU

The alliance between Sun Microsystems and Fujitsu Microelectronics, Inc. (FMI), is an example of a successful U.S.-Japan computer/semiconductor company alliance that demonstrates how U.S. companies can leverage Japanese components, manufacturing, and distribution channels to establish worldwide market share. The Sun-Fujitsu strategic alliance has been crucial to Sun Microsystems' success in the highly competitive workstation market. FMI produces 32-bit RISC processors for Sun workstations based on Sun's SPARC (Scalable Processor ARChitecture) operating system and distributes Sun workstations throughout Asia. Fujitsu may also become a second source for SPARC-based laptop workstations, reducing Sun's current sourcing of Toshiba laptops.

The alliance has delivered benefits to both companies. It has established Sun as the leader in the fast-growing workstation industry. Sun achieved a half-billion dollars in sales in just five years and has captured nearly 25 percent of the Japanese workstation market. Without FMI's early commitment, Sun might not have distinguished itself as an industry leader. The alliance enabled FMI to enter the laptop workstation market. Among workstation users, SPARC has become a de facto industry standard. To consolidate its market position, Sun has established SPARC International with FMI and 38 other U.S., Asian, European, and Japanese computer and chip makers (see Figure 17).

Ironically, Fujitsu was not Sun's first choice for a partner. According to William N. Joy, a Sun founder and R&D vice president, Sun approached most large domestic (U.S.) semiconductor companies—more than 20 in all—about a possible partnership in SPARC architecture. For a variety of reasons, none of the U.S. semiconductor makers were willing or able to make the $3 million to $4 million investment required to develop the SPARC chip.

Wayne Rosing, Sun's vice president for high-end engineering, explained that Sun needed a single chip (exclusive of floating point) capable of running at 10 MIPS (million instructions per second) for its Sun-4 workstation. In 1983, the year of Sun's search, a 20,000-gate array (semicustom chip) with 4,906 bits of register file memory in static random access memory needed for the task was not commercially available from any U.S. chip maker. LSI Logic was selling 10,000-gate arrays and had designed a 50,000-gate array, but it would not ship silicon until 1986—too long for Sun to wait in the highly competitive workstation market. A standard U.S. chip supplier would have had to commit significant custom engineering sources to the job, which was difficult during the high-growth semiconductor and computer markets in 1983-1984. Instead, U.S. chip makers such as Intel, Motorola, and National Semiconductor tried to steer Sun to their standard

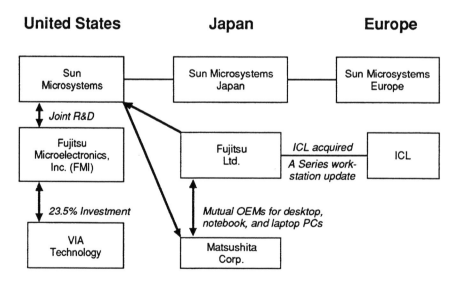

FIGURE 17 The Sun-Fujitsu SPARC alliance. SOURCE: Compiled by NRC Semiconductor Working Group.

microprocessor lines since the SPARC would take them out of their mainstream business. Advanced Micro Devices, Fairchild, and Texas Instruments had technical solutions, but they were six to twelve months behind Sun's schedule. Fairchild had a 5-MIPS Clipper processor, but it was too slow. Thus, Sun could not find a suitable U.S. partner.

For their part, U.S. companies considered Sun's overtures too risky. Sun was an unknown start-up company among many potentially viable competitors. Although reduced instruction set computer (RISC) was attracting attention, it was an unproven commercial technology. Microprocessor vendors preferred to focus on their own complex instruction set computer (CISC) microprocessors.

In a last-ditch effort, Sun went to FMI's headquarters in Silicon Valley to secure a microprocessor manufacturer. According to Ken Katashiba, general manager of FMI, Sun's timing was fortuitous since FMI was seeking to invest in emerging technologies and growth companies. Katashiba felt that producing SPARC chips for Sun could position FMI in the high-growth technical workstation market. However, convincing Fujitsu's top management in Tokyo, which Katashiba described as having a "mainframe mentality," was difficult. However, executive vice president and director, Matami Yasufuku, founder of Fujitsu's semiconductor operations, liked Sun's proposal. FMI had a 20,000-gate array being considered by two minisupercomputer start-ups, Convex Computer and Alliant Computer, which were using FMI's 8,000-gate arrays. According to Sun's Bill Joy, "I don't think they [Fujitsu] spent days or months agonizing over it. That's perhaps my frustration with a lot of American companies. They spend so much time making up their minds that they miss the good opportunities. They don't act quickly enough."[43]

After committing engineers to the task in 1984, FMI began producing S-16, a 10-MIPS SPARC microprocessor, for the Sun-4 workstation in 1985. Sun workstations became a hit in the marketplace, enabling it to take the lead worldwide with its Sun-4.

The Sun-Fujitsu alliance has evolved over time. In 1988, Sun FMI and Wind River signed an agreement to accelerate the use of SPARC in real-time computing markets. Fujitsu also signed a five-year agreement to market Sun workstations in Japan, which boosted Sun to the leading position, ahead of Sony Microsystems' NEWS workstations and Hewlett-Packard's machines. Table 15 contains a brief chronology of highlights in the Sun-Fujitsu alliance.

During this period, Sun had the opportunity to diversify its SPARC chip sourcing beyond FMI. Essentially, Sun had four options: (1) stay with

---

[43] Norm Alster, "How Intel and Motorola Missed the Sun Rise," *Electronic Business*, November, 1987.

TABLE 15 Evolution of the Sun-Fujitsu Alliance

Date	Joint Development Activity
1986	Fujitsu develops the S-16 SPARC central processing unit (CPU) for Sun workstations.
June 1988	Fujitsu, Sun, and Wind River Systems announce a cooperative R&D effort to accelerate use of SPARC architecture in real-time computing markets.
August 1988	Sun and Fujitsu sign a five-year OEM agreement valued at $280 million. Fujitsu begins marketing Sun workstation servers (S family) for the Japanese market.
January 1991	Through its newly acquired British subsidiary, ICL, Fujitsu reconfigures its 68000-based "A series" around the SPARC chip and UNIX System V, Release 4 operating system.
May 1991	FMI develops two 32-bit RISC chip sets named SPARClite^{TM} (MB869308, MB86940) based on SPARC architecture, which it began shipping in July 1991. These chips are targeted for the embedded control rather than the workstation market.

FMI, (2) make its own SPARC chips, (3) find a second source, or (4) establish a SPARC consortium. Table 16 lists the trade-offs of these options. Sun could manufacture its own chips—the classical "make or buy" decision—but semiconductor manufacturing is a totally different business that would require heavy upfront investments. Going it alone is a very risky strategy, especially because Sun had no semiconductor experience. Moreover, RISC is becoming a commodity market as more chip makers enter the market. Why didn't Sun choose a second source in the United States to reduce its dependence on Fujitsu? Sun could have found a second source, but it decided to stay with Fujitsu, which invested an additional $280 million for a five-year exclusive distribution agreement in Asia. This is much more help than a U.S. chip maker could have offered. It gave Sun not only financing for next-generation SPARC development, but also access to the fast-growing Asian market. However, to reduce its dependency on Fujitsu, Sun has expanded its SPARC International consortium to 39 members. LSI Logic, ROSS Technology (Cypress subsidiary), Texas Instruments, and Weitek are U.S. members. This option gives Sun the advantage of building on the Fujitsu relationship, while consolidating its market position and diversifying its chip source.

Sun's alliance with FMI extends beyond Silicon Valley, where FMI's Advanced Products Division works closely with Sun to develop next-generation RISC processors. FMI owns 23 percent share of a start-up company, VIA Technologies, which is developing RISC processors for computer peripherals. In Europe, Fujitsu Ltd. recently acquired the British computer maker ICL, which will remodel Fujitsu's A series workstations using the SPARC operating system. Sun is not dependent solely on Fujitsu. Sun supplies workstations to Matsushita and Oki, which sell them under their own labels. Fujitsu and Matsushita have close computer ties. Toshiba uses the SPARC operating system for its laptop workstations. While these alliances put Japanese companies into the workstation market, they help Sun against its immediate competitors: Hewlett-Packard, IBM, Sony Microsystems, and MIPS Computer (ACE).

Sun is entering a variety of strategic alliances with Japanese companies to position its workstation in the Japanese market. Besides supplying SPARC workstations to Toshiba, Oki Electric, and Matsushita, Sun Microsystems Japan is working with Morisawa on font development and negotiating with the Taiwanese vendors Tatung and DTK Computer, which want to sell Sun-compatible workstations in Japan. Early in 1991, Sun began negotiating separately with Fujitsu, Toshiba, and Matsushita to develop the next-generation multimedia RISC workstations by using their digital video data base, video processing integrated circuits, video windowing, and integrated services digital network (ISDN) technologies. In addition, Sun has an extensive network of 2,000 value-added resellers (VARs) worldwide to promote hardware sales and software development agreements to develop applica-

TABLE 16 Potential Tradeoffs of Sun-Fujitsu Alliance for Sun

	Advantages	Disadvantages
FMI only	Reliability, stable source $280 million deal	Dependence Potential competitor
Build chips	Control over technology Leverage over Fujitsu	Heavy investment No semiconductor know-how Unsure supply
Second source	Lower prices Leverage over Fujitsu	Administrative time Undercut Fujitsu alliance
Consortium	Lower prices Diversified chip supply Consolidate SPARC position	Administrative time Need to educate members

tions packages. Since MIPS Computer announced its Advanced Computing Environment (ACE) consortium, Sun is expanding SPARC International. Table 17 lists Asian members of Sun's SPARC alliance.

Sun's strategic alliances are designed to improve its position in the Japanese market. The alliances range from RISC processor development (Fujitsu) and supply agreements (Toshiba, Oki, Matsushita, Tatung, DTK) to font development (Morisawa) and multimedia technologies (Fujitsu, Toshiba, Matsushita). The Sun-Fujitsu model provides the foundation for both these and the other SPARC alliances.

The Sun-Fujitsu alliance has created a "domino effect" among its VARs and end users, which have entered other alliances. Oki Electric, for example, uses Intel's 80860 microprocessor in its Oki Station 7300 workstation, which is supplied to U.S. minisupercomputer maker Alliant Computer. Toshiba, which supplies SPARC laptops to Sun, is planning to use the R4000 microprocessor from Sun's rival, MIPS Computer. Table 18 summarizes these activities.

The major impact of the Sun-Fujitsu alliance can be seen in the emergence of rival RISC computer consortia, as shown in Table 18. MIPS Computer, which completed a two-year exclusive agreement with NEC for its R3000 workstations, is working with ACE to compete with the Sun workstation standard.

TABLE 17   Sun's Japanese SPARC Alliances

Japanese Company	Joint Activity
Fujitsu Microelectronics, Inc. (FMI)	Responsible for developing and producing SPARC chips for Sun
Toshiba	Procures SPARC processor from Sun for its laptop workstation
Matsushita Computer Systems	Marketing Sun workstations under its own brand name
Morisawa	Technical tie-up for workstation font development
Tatung-DTK Computer Japan	Sun negotiating over Japanese version of Sun operating system with both companies, which plan to ship Sun-compatible workstations
Fujitsu, Toshiba, Matsushita Electric	Sun negotiating separately with Japanese companies to develop multimedia workstation technologies for ISDN

TABLE 18 SPARC Workstation Alliances, 1991

Vendor	Partner	Agreement
Fujitsu	VIA Technologies	23.5% investment in VIA to develop RISC MPUs
Fujitsu	Sun Microsystems	Fujitsu produces SPARC chip, while Sun OEM supplies workstation to Fujitsu and Toshiba
Fujitsu	ICL	Developing SPARC-based UNIX workstation
Matsushita Corp.	VIA Technologies	4.7% investment in VIA to develop RISC MPUs
Matsushita Electric	Sun Microsystems	Sun OEM supplies workstations to Matsushita
Mitsubishi Electric	VIA Technologies	Owns 4.7% of VIA; distribution via Diasemicon Systems
Oki Electric	Sun Microsystems	Sun OEM supplies desktop workstations to Oki
Oki Electric	Stardent Computer	Oki OEM supplies Oki Station 7300 workstation
Oki Electric	Alliant Computer	Oki OEM supplies Oki Station 7300 workstation
Toshiba	Sun Microsystems	Toshiba OEM supplying RISC laptops to Sun, while Sun OEM supplies desktop workstations to Toshiba
Toshiba	LSI Logic, Siemens	Planning to make R4000 for LSI Logic and Siemens
Toshiba	Unisys Japan	Supplying laptop workstations to Unisys Japan
Unisys Japan	Sun Microsystems	Sun OEM supplies desktop workstations to Unisys

ACE signals a trend toward global RISC workstation consortia. Recently, Hewlett-Packard licensed its Precision Architecture to Hitachi, Mitsubishi, and Samsung, which will supply workstations on an OEM basis to Hewlett-Packard. Intel Japan has teamed up with six companies in the United States and Japan to promote its i860 RISC processors. Motorola has the "88 OPEN" consortia for its 88000 RISC processor, and it will supply chips to IBM and Apple Computer for their RS/6000 workstations. In several years, all major workstation makers will belong to a RISC consortium. Sun has no choice but to expand beyond its Fujitsu alliance to SPARC International.

The significance of this shift from binational alliances to global consortia reflects the trend toward open computer architectures and more competition. Sun and MIPS Computer have successfully shown that by aggressively licensing their RISC technology to Japanese and other companies, they can establish their operating systems as industry standards. Fear of losing sales to one's partner is greatly overshadowed by the fear of becoming obsolete in terms of industry standards. IBM, for example, recently dropped its go-it-alone strategy for its RS/6000 workstation, which has done poorly in the marketplace, and teamed up with Motorola and its potential rival Apple Computer. In personal computers, IBM has Matsushita manufacture its PS/2 and PS/55 in Malaysia. For U.S. workstation makers, the long-term risks of creating one's competitors is overshadowed by the more immediate threat of being eliminated altogether from the market. Although this is not an enviable choice, all major U.S. workstation vendors are pursuing a global consortium strategy to maintain competitiveness.

Thus, the computer industry is likely to see a proliferation of strategic alliances in the incipient R&D phase, which gradually expand into global consortia when the technology has carved out a market niche. U.S.-Japan alliances are becoming increasingly important because Japanese companies dominate many leading-edge technologies, such as LCDs, memory chips, IC cards, thin packaging, and mass manufacturing. The alternative for U.S. computer companies is not whether they should go it alone or team up only with U.S. companies, which may not have the critical components, but how best to take advantage of Japanese component strengths to expand their worldwide market share.

What are the lessons to be drawn from the Sun-Fujitsu alliance? The alliance demonstrates that it is possible for U.S. companies to leverage a Japanese partner into expanded worldwide market share. Although this dependence on Japanese companies may not be the ideal situation, it is clearly better than not being in business at all or remaining a minor player.

As Sun's Bill Joy admits, there is no such thing as a "free lunch." Fujitsu might become a competitor in the workstation market. In fact, Fujitsu has invested $40 million in Hal Computer Systems, which is headed by an ex-IBM workstation manager, to develop a 64-bit superscalar microprocessor based on the SPARC architecture. For U.S. companies, there are always major trade-offs that must be considered when entering alliances with Japanese companies, as shown in Table 19. Only by considering these trade-offs can U.S. companies identify the long-term dangers. On the other hand, the Sun case demonstrates that remaining too conservative or slow-footed also means missing market opportunities. Fortunately, Japanese partners are no longer the only option for U.S. companies. The rise of global consortia may reduce U.S. dependence on Japanese alliances by broadening the sources of components, manufacturing, and distribution.

TABLE 19 Trade-offs of the Sun Microsystems-Fujitsu Alliance

Benefits	Costs
Access to engineering support	Potential loss of technology
Access to "patient money"	Vulnerable to outside pressure
Technology licensing to set standard	Create new competitors
Leverage partner's global organization	Time consuming and administrative costs
Access to manufacturing capacity	Loss of ability to manufacture
Joint marketing, sales, and distribution	Potential conflict over markets

However, the effectiveness of these consortia remains to be seen because they are difficult to coordinate and manage. Moreover, the decline in American competitiveness in critical components and the unwillingness of U.S. chip makers to divert from their mainstream businesses may still force U.S. computer makers to seek Japanese partners as a shortcut to global market success.

CASE III: KUBOTA COMPUTER

Kubota Computer is a classic example of a large Japanese company acquiring U.S. technology in order to diversify into high-growth markets. The heavy farm machinery company had been severely affected by the "yen shock" of 1986 and liberalization of agricultural imports. To offset stagnant and declining products, Kubota has systematically invested in a variety of emerging technologies critical to next-generation computers, ranging from workstations and minisupercomputers to magneto-optic disk drives and software, as shown in the Table 20. Kubota is following other Japanese steel and chemical companies that are quickly diversifying into high technology, including Asahi Chemical, Kobe Steel, Nippon Steel, Nippon Kokan, Kawasaki Steel, and others. However, what makes Kubota unique are the breadth and speed of its acquisitions.

In mid-1986, Kubota began by investing $19 million in Ardent Computer, a minisupercomputer maker, which later merged with Stellar Computers to form Stardent Computers. Both Ardent and Stellar had leading-edge technologies but faced financial difficulties when sales were slower than expected. In exchange for a 22 percent equity share, Kubota agreed to market Stardent's TITAN minisupercomputers in Japan. Kubota also signed a contract to have Nippon Steel market the TITAN computer. The Kubota-Stardent partnership had four parts:

1. Kubota held a 22 percent equity share and had infused $55 million after the merger.

TABLE 20 Kubota's Acquisitions and Investments

Company	Business	Date	Investment ($ millions)	Share (%)
Stardent Computer	Minisupercomputers	7/86 8/88 9/89	19 25 24	22
MIPS Computer	RISC workstations	10/87 1/90	25 10	18
Akashic Memories	Thin-film media	12/87 1/88	16 10	100
Maxtor/Maxoptics	Magneto-optic drives	3/89	12	25
Rasna	Mechanical CAD and software	6/89	10	15
Exabyte	Computer magnetic storage	—	0	Board representation
C-Cube Microsystems	Image processing video compression	3/90	6.25	39.5
Teknocom	CAD software	1/91	0.23	37.5
Allied Information	Software	6/91	—	—
Tricord Systems	Computer servers	6/91	3.4	—

2. Kubota had sole marketing rights in the Pacific Rim.
3. Kubota did all manufacturing and set prices to Stardent.
4. Kubota had rights to all Stardent technology.

In essence, Stardent gave away its technology and high-growth Asian markets for continued existence as an R&D laboratory. In the autumn of 1991, Stardent closed down, wiping out Kubota's considerable investment.[44] Kubota Pacific Computer took over the TITAN line, Stardent's biggest operation.

At about the time the alliance with Stardent was launched, Kubota acquired 18 percent of MIPS Computer, a fast-growing RISC workstation maker in Silicon Valley. Kubota's goal is to resell MIPS workstations in

---

[44]William M. Bulkeley and Udayan Gupta, "Japanese Find U.S. High Tech a Risky Venture," *Wall Street Journal*, November 18, 1991, p. B1.

Japan and eventually develop its own proprietary workstations based on MIPS' R3000 and R4000 microprocessors. In 1990, newly created Kubota Computer and MIPS Computer jointly developed the RC6280, a minisupercomputer that runs the Unix operating system, a high-growth area in Japan. Kubota is supplying the workstation to MIPS Computer on an OEM basis. Kubota also has exclusive rights to manufacture and produce the MIPS RS3230 workstation in Japan. To achieve its sales goal of $100 million, Kubota created an independent workstation department, organized a distribution network, and doubled its MIPS workstation sales and technical staff.

The Kubota-MIPS Computer relationship has deepened in recent years. In 1991, Kubota was commissioned by MIPS Computer to produce RISC workstations in order to reduce production and distribution costs and shorten delivery time for MIPS products in Japan and Asia. In mid-1990, Kubota joined MIPS' ACE consortium, whose members will develop advanced 64-bit RISC workstations based on the MIPS R4000 microprocessor.

In addition to computers, Kubota has invested in Akashic Memories (thin-film media), Maxtor/Maxoptics (magneto-optic disk drives), C-Cube Microsystems (image processing and video compression), Teknocom (CAD software), Rasna (mechanical CAD and analysis software), and Allied Information (software). Allied Information Systems, a merger of software firms Dynatech International, Applicationware Research, Conam, and World Business Wing, will enable Kubota and Allied to set up a distributed development system and establish regional R&D centers to hire university graduates in software programming. The emerging hardware and software technologies acquired through this diversification will position Kubota Computer in advanced workstation and minisupercomputer markets.

Like other foreign companies, Kubota hires U.S. attorneys, merger and acquisition experts, and management consultants to assist in identifying and negotiating its acquisitions and investments. Indeed, the large number of U.S. high-tech bankruptcies, lack of venture capital, and long payback periods have created a large market for "technology brokers" who sell off U.S. companies that are technologically rich but financially poor. This development is an extension of the large merger and acquisition activities of the 1980s but is a shift to high-technology assets and global buyers because the U.S. merger and acquisition market has declined sharply. Due to the overall computer industry slowdown and the insufficiency of U.S. venture capital, the selling of U.S. high-tech companies may accelerate during the 1990s.

What are the lessons of Kubota for U.S. companies? Japanese acquisitions are a mixed blessing. Besides giving away technology, they are vulnerable to lawsuits and management problems. In mid-1990, Kubota Computer became embroiled in a lawsuit alleging that Kubota had seized control of Stardent to secure its technology. Stardent fired its cochairmen, Allen Michels and Matthew Sanders, after they filed a $25 million lawsuit. A

Kubota board director who investigated the suit also resigned. The lawsuit was complicated because of Kubota's contention that both cochairmen offered to resign from Stardent, sell their stock, and not bring any lawsuits in return for five-year consulting contracts worth $3.5 million each. Kubota apparently rejected the offer. The lawsuit and Stardent's subsequent failure may have a chilling effect on Japanese investments and acquisitions in U.S. high-tech companies.

The failure of Stardent shows that investments in U.S. high-technology companies do not always pay off for Japanese companies, but Kubota's goals and expectations for its U.S. alliances go beyond short-term financial considerations. Alliances are designed to help Kubota build new technical competences in support of an integrated diversification strategy. Even when ventures fail, Kubota and other Japanese companies pursuing similar strategies retain the competence and ability to use the technology.

Appendix B

Examples of Japanese Acquisitions and Investments in U.S. Semiconductor Companies

Japanese Investor	U.S. Company	Year	Product/ Technology	Equity (% or $millions)
1. ASCII	Nexgen Microsystems	1990	microprocessors	$5/100%
2. ASCII	Tera Micro Systems	1990	peripheral chips	$2
3. ASCII	Crosspoint Solutions		gate arrays	50%
4. Fujitsu	Hal Computer	1991	microprocessors	$40.2/44%
5. Fujitsu Microelectronics	VIA Technologies	1989	IC chips	23.5%
6. Hitachi Kasei	Kollmorgen R&D Ctr.		circuit boards	100%
7. Hitachi, Ltd.	Ceraclad		ceramic packages	100%
8. Ishihara Sangyo Kaisha	Mountain View Research	1990	circuit boards	$6/100%
9. Koito	North American Lighting			$16
10. Kubota	C-Cube Microsystems		graphics VLSI	36.1%
11. Kubota	MIPS		RISC	17.8%
12. Matsushita	National Semiconductor	1990	manufacturing plant	$86/Division
13. Mitsubishi Corp.	VIA Technologies	1990	MPUs	$1
14. Mitsubishi Corp.	Tera Microsystems	1991	RISC chips	$.8/6.7%
15. Mitsubishi Electric	Powerex		discrete devices	33%
16. Mitsui Comtek	Integrated CMO Systems		ASICs	
17. Mitsui Petrochemical	Kodak (Pathtek)	1990	3-D devices	$4/Division
18. Nikko Capital Co.	DSP Group	1991	digital chips	$7
19. Nikon Metals	Anacomp	1989	thin film disc	Division
20. Nippon Steel	Simtek		semiconductors	$2
21. NKK	Paradigm Technology	1991	SRAMs	$5

continued on next page

APPENDIX B *Continued*

Japanese Investor	U.S. Company	Year	Product/ Technology	Equity (% or $millions)
22. Oki Electric	Vitelic		ICs	
23. Olin Asahi	Aegis		IC packages	100%
24. Ono Sokki	KLASIC	1991	circuit boards	$4.3/Division
25. Otari Electric	King Instrument	1990	tape loaders	100%
26. Ozaki Elec.	Planade Energy Systems		power meters	100%
27. Ricoh	Panatech Research		semiconductors	Division
28. Rohm	Xetel	1990		$3.2
29. Sanken	Sprague Technology	1990	semiconductors	$58/Division
30. Shin-Etsu Chemical	Brooktree		graphic chips	$2
31. Shin-Etsu Semiconductor	Cree Research		blue LEDs	$2
32. Sony	Advanced Micro Devices	1990	manufacturing plant	$55/Division
33. Sumitomo Metals	Mosaic Systems	1990	multi-chip modules	$2
34. Sumitomo Metals	nCHIP	1991	multi-chip modules	
35. TDK	Silicon Systems	1989	ICs	$200
36. Toko	Signal Processing Technology		digital image chips	100%
37. Tokuyama Soda	Allegro Microsystems	1990	semiconductors	$58/Division
38. Tokuyama Soda	General Ceramics		ceramic package	$59
39. Toppan Printing	Texas Instruments	1990	photomasks	$19/Division
40. Toppan Printing	Prostar	1990	circuit boards	100%
41. Toshiba	Integrated CMO Systems		ASICs	
42. Toshiba	Synergy Semiconductor	1991	SRAMs	
43. Toshiba	Vertex Semiconductor	1991	ASICs	$20
44. Uemura	Automated Semiconductor	1989	IC plating	$.4
45. Yamaha	Sequential Circuits		ICs	

NOTE: List does not include the significant "green field" investments in the United States by the major Japanese semiconductor companies.

SOURCE: Compiled by OJA Staff from data provided by the Economic Strategy Institute and the American Electronics Association Japan Office.

Appendix C

Examples of Japanese Acqusitions and Investments in U.S. Semiconductor Equipment and Materials Companies

Japanese Investor	U.S. Company	Year	Product/ Technology	Equity (% or $millions)
1. Advantest	Sym-Tek Systems	1988	test equipment	
2. Canon	AG Associates		semiconductor eqpt.	$3m
3. Canon	Cymer Laser Technologies	1990	eximer lasers	
4. Canon	Lepton	1990	masking	49%
5. Dainippon Screen	MRS		semiconductor eqpt.	
6. Fuji Electric	PPC Industries		electron beam	
7. Fujikin	Carten/Marten	1989	valves	
8. Ebara Corp.	Varian Associates	1991	cryopump div.	$7.5m/Division
9. Hamamatsu	Inspex		semiconductor eqpt.	100%
10. Hoya	Micromask Inc.	1989	photoplates	$25.3/100%
11. Ishikawajima-Harima Heavy Industries	Siscan Systems	1990	contamination eqpt.	100%
12. Iwasaki Electric	Energy Sciences	1988	electron beam	100%
13. Kobe Steel	GCA Laser Inspection		laser inspection	100%/Division
14. Kokusai Elec.	BTU International	1992	diffusion furnace	$24/Division
15. Komatsu	Union Carbide	1990	polysilicon	Division
16. Kyocera	AVX	1990	ceramic capacitors	$575
17. Marubeni Hytech	Mattson Technology Inc.	1991	wafer fabrication	
18. Mitsubishi Corp.	Cymer Laser Technologies	1991	eximer lasers	$2
19. Mitsubishi Metals	Siltec		silicon wafers	100%
20. Mitsui & Co.	Ergenics	1989	vacuum getters	

continued on next page

APPENDIX C *Continued*

Japanese Investor	U.S. Company	Year	Product/ Technology	Equity (% or $millions)
21. Nikon	Cymer Laser Technologies		eximer lasers	
22. Nippon Gaishi	Cabot (Beryllium)		materials	Division
23. Nippon Kokan	General Electric		silicon factory	Division
24. Nippon Mining	Gould	1988	copper foil	$1,100
25. Nippon Sanso	Matheson Gas Products	1989	semiconductor gas	100%
26. Nippon Sanso	Semi-Gas Systems	1990	gas handling	$23
27. Nippon Sanso	Tri Gas Inc.	1992	semiconductor gases	$88/100%
28. Nippon Steel	Holon			
29. Nitto Boseki	Midland Bioproducts		semiconductor eqpt.	
30. Osaka Titanium	Cincinnati Milacron	1989	materials	Division
31. Osaka Titanium	U.S. Semiconductor		wafers	100%
32. Seki & Co.	Novellus Systems	1988	front-end	$1.9
33. Shin-Etsu Chem.	MicroSi		silicon	100%
34. Sony Corp.	Materials Research Corp.		wafer process	100%
35. Sumitomo Corp.	LTX Corp.		semiconductor eqpt.	30%
36. Sumitomo Corp.	Prometrix	1991	measuring eqpt.	$1m
37. Sumitomo Heavy Industries	Radiation Dynamics		e-beam	
38. Sumitomo Metals	APT		semiconductor eqpt.	5%
39. Sumitomo Metals	Lam Research		etching, front-end	$5/4.5%
40. Sumitomo Metals	LTX Corp.	1990	testing equipment	$24m
41. Toray+ Shimadzu	Therma Wave		semiconductor eqpt.	65%(Toray) 22%(Shimadzu)
42. Toshiba Ceramic	Quartz Industrial		materials	80%
43. Tosoh	Varian Associates	1989	materials	$33m/Division
44. Tosoh	Weiss Scientific Glass Blowing	1990	glass for IC fab	
45. ULVAC Japan	BTU-ULVAC	1991	wafer processing	

SOURCE: Compiled by OJA Staff from data provided by the Economic Strategy Institute and the American Electronics Association Japan Office.

Appendix D

Workshop on U.S.-Japan Technology Linkages in Semiconductors: Agenda and Participants

September 12, 1991
Stanford University - Encina Hall

National Research Council
Committee on Japan

Welcome and Introductions

Paper Summary
DANIEL OKIMOTO, Stanford University

Historical Experience/Trends
SHERIDAN TATSUNO, NeoConcepts
Discussants:
Warren E. Davis, Semiconductor Industry Association
Bevan P.F. Wu, IBM

Break

A Japanese Perspective
YOSHIO NISHI, Hewlett-Packard
Discussants:
Masazumi Ishii, AZCA
Hideo Ito, Toshiba America Electronic Components
Tai Sato, Kubota C-Cube Microsystems

Case Studies
EDWARD DEWATH
Discussant:
John A. Blair, Booz Allen & Hamilton

Discussion of Paper
WILLIAM HOWARD, National Academy of Engineering

Other Participants/Discussants
  Elie Antoun, LSI Logic Corp.
  Michael Borrus, University of California, Berkeley
  Robert Burmeister, Saratoga Technology Associates
  Papken Der Torossian, Silicon Valley Group, Inc.
  Kathleen Eisenhardt, Stanford University
  William Lane, Motorola Inc.
  Junko Matsubara, Dataquest Inc.
  Franklin Schellenberg, IBM